I0895812

Mosquito Special Operations in the Second World War

Mosquito Special Operations in the Second World War

The Ultimate Fighter Bomber

Peter Saxton

Pen & Sword
AVIATION

First published in Great Britain in 2024 by
Pen & Sword Aviation
An imprint of Pen & Sword Books Limited
Yorkshire – Philadelphia

Copyright © Peter Saxton 2024

ISBN 978 1 39905 948 0

The right of Peter Saxton to be identified as
Author of this Work has been asserted by him in accordance
with the Copyright, Designs and Patents Act 1988.

A CIP catalogue record for this book is
available from the British Library

All rights reserved. No part of this book may be reproduced or
transmitted in any form or by any means, electronic or mechanical
including photocopying, recording or by any information storage and
retrieval system, without permission from the Publisher in writing.

Typeset by Mac Style
Printed in the UK by CPI Group (UK) Ltd, Croydon, CR0 4YY.

MIX
Paper | Supporting responsible forestry
FSC® C013604

Pen & Sword Books Limited incorporates the imprints of After
the Battle, Atlas, Archaeology, Aviation, Discovery, Family History,
Fiction, History, Maritime, Military, Military Classics, Politics,
Select, Transport, True Crime, Air World, Frontline Publishing, Leo
Cooper, Remember When, Seaforth Publishing, The Praetorian Press,
Wharncliffe Local History, Wharncliffe Transport, Wharncliffe True
Crime and White Owl.

For a complete list of Pen & Sword titles please contact

PEN & SWORD BOOKS LIMITED
47 Church Street, Barnsley, South Yorkshire, S70 2AS, England
E-mail: enquiries@pen-and-sword.co.uk
Website: www.pen-and-sword.co.uk
or
PEN AND SWORD BOOKS
1950 Lawrence Rd, Havertown, PA 19083, USA
E-mail: uspen-and-sword@casematepublishers.com
Website: www.penandswordbooks.com

Dedicated to my wife Barbara

Contents

Chapter 1

Introduction

In September 1941, a single unarmed aircraft roared off the runway of RAF (Royal Air Force) Station Benson in Oxfordshire. This was a reconnaissance aircraft and Benson, their base, was conveniently close to RAF Medmenham. RAF Medmenham was the air photo-reconnaissance interpretation intelligence centre for the British armed forces and their allies.

The photo-reconnaissance interpreters comprised a high proportion women of the Women's Royal Navy Service (WRNS) and Women's Auxiliary Air Service (WAAFS) who were proving themselves particularly clever at photo interpretation for targeting bomber strikes.

The two crew members of this aircraft settled themselves and their two engines to the task. They began a long climb towards their planned cruising altitude above 30,000 feet.

The aircraft was picked up on German radar. A flight of Messerschmitt 109 fighter aircraft was scrambled and vectored to intercept. The Luftwaffe pilots, experienced veterans who knew their business, eventually made visual contact with a strange light blue twin-engined aircraft above them. As they pulled up, switched on their guns, and tried to position for an attack, their aircraft started to judder; they were close to the 109's altitude performance ceiling and they weren't going to make it. Their quarry, still well out of reach above them, pulled away from them at surprising speed making a chasing interception impossible.

In September 1942, N105 Squadron RAF, bombers of Royal Air Force Bomber Command, crossed the North Sea at low-level to avoid detection. They carried out a roof-top level attack on the Gestapo headquarters in Oslo; and destroyed it.

In October 1942, a night fighter of RAF Fighter Command was vectored by WAAFs in ground radar control, on a black moonless night, onto and behind a Luftwaffe Junkers 88 medium bomber. In the vibrating roaring dark, the navigator, head pressed hard into the radar scope said.

> 'I've got him. Rudder her left five degrees. He's in front now – 400 yards – slightly above.'

The pilot, head down glued on instruments in a darkened cockpit, without daring to move his head, moved the gun master switch to FIRE and placed his gloved fore-finger over the cannon firing trigger. He took a breath and still without moving his head, glanced his eyes up and ahead. There he saw the glow of exhausts. Allowing the plane to drift upwards, he closed in to 250 yards and squeezed the cannon trigger.

In March 1943 an aircraft of RAF Fighter Command crossed the North Sea and into enemy occupied Europe keeping low until he had coasted in. They acquired their target area, which surrounded a Luftwaffe aerodrome. Levelling at 5,000 feet, they throttled back, trimmed the aircraft, checked their fuel endurance and waited. This crew was on an Intruder mission. The aircraft kept its distance, the crew patiently watching. The setting sun was low in the sky, and the aircraft, up-sun of the aerodrome and in and out of cloud, was invisible and inaudible to those at the aerodrome.

Eventually the Navigator said quietly, 'There's an aircraft two o'clock low – there – and he has his landing lights on'. The pilot set up a descending circuit and dropped down behind the aircraft. And opened fire.

In March 1945, fighter-bomber aircraft from RAF Coastal Command carried out devastating rocket attacks on shipping in the Skagerrak. One of those aircraft was hit by flak. The crew shut down the port engine, and then flew back across the North Sea 'on one' to land safely at RAF Dyce near Aberdeen in Scotland.

It is impressive that by this time the Herculean efforts of the British aircraft industry had been able to supply the RAF with a range of aircraft types wide enough to enable its squadrons to carry out all these tasks and more. What is more impressive though, is that the very differing missions referred to above were all carried out by just one aircraft type, an aircraft of such innovative design and such strategic import as to be without peer in mission flexibility, effectiveness, and cost-effectiveness. This aircraft began its career as the de Havilland prototype DH98.

It was to become legendary as the The Mosquito.

Chapter 2

Strategy

The Royal Air Force was shaped between the wars as a bomber force. In an eerie precursor to the Cold War, many in the Great Powers accepted that, as Stanley Baldwin, a proponent of the Italian strategy theorist Giulio Douhet, had stated to The House of Commons in 1932:

'…the bomber will always get through…'

While this could and was taken by some as an argument for disarmament, there were others who inferred a contrary strategic interpretation; that sovereign states should arm themselves with bomber fleets capable of inflicting unacceptable damage on any potential enemy. It was the age-old strategy of deterrence. This came to be appropriately named MAD, short for 'Mutually Assured Destruction', in the later, nuclear-armed Cold War.

The Cold War was a thundering echo of the sort of dilemmas that had been besetting strategic theorists and military planners earlier in the 1930s. In the 1950s there emerged an unimaginably more dangerous situation. The strategic elements such as arms races and deterrence were still present, but this time it was brought into sharp focus because both sides were armed with nuclear weapons. It remained clear that the optimum outcome in a nuclear bi-polar world would have been for both sides to give up all nuclear weapons and indeed many strenuous attempts were made. The cost of the continuing development, manufacture, servicing and deployment was astronomical on both sides. So was the cost of the development, manufacture and upkeep of the required stockpiles; and of delivery systems. Strategic bombers such as the Boeing B52 at first; and later Inter-Continental Ballistic Missiles (ICBM). But the international strategic system critically lacked trust – you had to do it yourself because you could never be sure that the other side would sign a disarmament treaty and then cheat by arming anyway. This was the most famous of all Game Theory scenarios, known as 'The Prisoner's Dilemma'. Instead of by disarming, the face-off gained stability from the mutually shared notion that any attack on another state would inevitably bring disastrous retaliation upon the aggressor; and so peace in our time. This thinking had clear roots going back to twenty years before. Its theoretical stability was marred, however, by two things of terrifying consequence – the possibility that one of the actors would not behave rationally; and the possibility of a simple, disastrous mistake.

In the 1930s the thinking about achieving a stable balance changed. German and British military thinkers realised that a strategic advantage might be gained by producing fleets of fast, nimble light air-fighting aircraft, capable of disrupting mass bomber formations by shooting them down in significant numbers. One's own cities might be saved from a first strike and total destruction by such fighter planes, while the ability to deliver retaliatory revenge through its bomber fleets remained intact. Now both sides of the scenario had to have both tactically defensive light air-fighters and strategically offensive bomber fleets to balance and re-stabilise the strategic equation. There can be little doubt that both Germany and Britain understood this, but notwithstanding, both nations got the elements of balance, if not wrong, less than optimum.

The light fighter escalation was both easy and necessary – they had to exist if the enemy had a bomber fleet; but what sort of a bomber fleet should a nation have? Here there was a difference in thinking between Britain and Germany. Germany opted for large fleets of medium bombers – twin engined machines such as the Junkers 88 and the Dornier 217. Germany did develop four-engine heavy bombers such as the Messerschmitt 264, and even a six-engine type – the Junkers 390 – but they achieved insufficient development and numbers to be deployed in the bombing campaigns of the Second World War. Germany continued with its preference for medium bombers.

Britain, however, while it deployed some medium bombers such as the Vickers Wellington, believed the future was still essentially with heavy bombers and went all out for heavy four-engined types such as the Short Sterling, the Handley Page Halifax, and the Avro Lancaster. If Germany lacked heavies, Britain's deficiency was in the medium range. Its twin-engined inventory was looking obsolescent with lessening operational effectiveness and inadequate performance especially in bomb load and range.

The Battle of Britain in 1940 seemed to confirm that the total annihilation of populations by fleets of attacking bombers as argued by General Douhet and described graphically by H.G.Wells in his novel *Things To Come*, had not yet come. Luftwaffe attacks, which had concentrated on RAF Fighter Command's airfields initially, as well as raids on London and other cities, did get through but at a grievous price in bombers shot down and crews killed or captured.

The construction of effective radar surveillance allowed RAF Fighter Command to dispense with the desperately wasteful system of standing air patrols. With standing patrols everything was hostage to fortune. There was the chance that when the enemy showed up you would be low on fuel and have to break away to refuel; the chance your aircraft would have become unserviceable on that sortie because of the increased overall rate of usage; the chance when the enemy showed up that you would be in completely the wrong position to attack; the chance you would be on the ground refuelling when they showed up unopposed and bombed your aerodrome; the chance you wouldn't see them in poor visibility; the chance you would be up there getting fatigued and they wouldn't show up at all.

Instead, pilots were held on the ground, their fighters fully refuelled and re-armed. Waiting. It can't have been very relaxing sitting reading an old copy of *Punch* under a Hurricane wing, but it was a lot more relaxing than flying it for hours to no effect. Also, you could take a nap – no chance of that while airborne. When the bell rang they were able to 'scramble' in minutes, be immediately vectored onto an intercept course and given an altitude at which to expect the opposition. The tallies of aircraft shot down mounted.

By the end of autumn 1941 the Luftwaffe, their losses deemed unsustainable, switched to night attacks. The RAF's single-engined, single-pilot fighters attempted to counter but with insufficient success; they were neither designed nor equipped for night fighting. Twin-engined two-crew Bristol Beaufighters rapidly modified with airborne radar were deployed. They began to shoot down bombers at night proving a concept. They showed the way.

Chapter 3

Industrial Matters

During the Great War and in the years between the two World Wars, aircraft manufacturing companies formed in the UK – many competing for government contracts. The normal process was to start with an Air Ministry specification and then brainstorm how the performance requirements could best be achieved. The designers would debate solutions, then get choices agreed and recommended with the board through the Chief Designer. De Havilland's were seeking the best way to produce a medium bomber that would fly at altitudes high enough and at cruise speeds so fast that fighter aircraft would be unable to intercept it. This implied that it would not need to be defensively armed, and this might have been where the concept started to come together. If it needed to go that fast and that high, it couldn't carry the burdens of heavy gun turrets and guns. But if it didn't have the burdens of heavy gun turrets and guns, it could, indeed, fly that fast and that high.

Why wood? Wood had been the material used in airframe construction until recently. But this most innovative of industries, in only a few years, had seen technologically driven improvements such as bi-planes replaced by monoplanes, from open cockpits to enclosed, fixed undercarriages replaced by retracting, from fixed pitch to variable pitch constant speed propellors, and in particular wooden aircraft being superseded by all-metal constructions. Once such a general transition occurred, past practices must have seemed *passé* and going back ill-advised, even perverse. However, there was a big problem about metal supplies. Metals of all kinds were a critical strategic material and in very short supply, so de Havilland's took another look at wooden construction as a way around a big problem that might have prejudiced the quantities of aircraft they could produce.

Wooden structures are easy to produce and cheap. They don't require expensive industrial facilities as does metal. They are light and can be made even lighter by manufacturing a lamination of balsa wood sheathed in plywood. Although it may be unintuitive, because it takes a greater volume of wood compared to metal to achieve the required structural strength, wooden structures withstand battle damage better; and holes in wood can be repaired more quickly, cheaply and easily than in metal.

Before the war, the conception, design and manufacture of military aircraft in Germany was dispersed. The Treaty of Versailles had prohibited Germany from having an air force at all, so aircraft were constructed secretly via a dispersed 'cottage' industry. Parts were manufactured in small plants and workshops all over the country, then transported to assembly points. These processes were difficult to detect by outsiders, and if detected, difficult to disrupt. They *were* detected however, and Winston Churchill announced what was going on to a startled House of Commons in 1937. He told The House, 'It's very clever!' Those words were noted by some key people, including, presumably, the remarkably savvy Sir Geoffrey de Havilland.

Wooden construction needed a workforce skilled in woodworking, and a way of recruiting and organizing it. Fortuitously, a major centre for the production of furniture was situated in Buckinghamshire not far from his aircraft factory at Hatfield in Hertfordshire. The furniture industry had been largely laid off idle, so here resided a skilled workforce that could manufacture aircraft parts precisely to blueprints. It too was a dispersed industry based on the 'cottage' and 'workshop' model – almost ideal. The various parts of wooden airframes, such as stringers and formers, could be contracted out to many small production units all over the area. Once completed they similarly could be transferred to a central assembly point and transformed into a completed aircraft. So as the German aircraft industry emerged from its covert, decentralized, manufacturing arrangements, de Havilland's embraced it with potentially all the security, organizational and skilling advantages Germany had employed to circumvent the restrictions of Versailles. This also manifested itself in the assembly process when the decision was taken to move the Mosquito work from the main factory at Hatfield to Salisbury Hall aerodrome some miles away. The factory at Hatfield was too obvious a target from the air, and indeed was attacked more than once. The most serious destroyed much of the work in progress, but luckily the three prototype Mosquitos remained unscathed.

The woods and forests of the south of England did offer raw materials such as birch and beech, but were more noted for deciduous forest, such as Burnham Beeches, than for evergreen pines. This was significant because the most important timber in aircraft structure was a particular kind of pine – the spruce. This timber formed the main 50-foot wing spars as well as other load-bearing components, but it was not so easily obtained from trees grown within the UK.

Three spruce forests near Chichester belonging to Lord Bathurst were a valuable source, supplying about six million cubic feet of timber of suitable quality. Despite that, it was not enough, so de Havilland's, a company as Canadian as it was British, sourced most of the rest of its requirements from North America. Typically, only 10% of trees passed muster as suitable for aircraft construction with grain that was straight and close and of sufficient age.

De Havilland's had started experimenting with wooden construction earlier in the decade with considerable success. They had produced the famous twin-engined Comet, which had won long distance air races. They had also produced a four-engined beautifully streamlined civil airliner, which they named 'Albatross'. They now started on designs for another twin-engined aircraft of the greatest aerodynamic efficiency producing minimum drag in flight. They wanted to incorporate the most powerful engines available, to achieve, in combination with the lightness of wooden fuselage and wings, the highest power to weight ratio possible. De Havilland's chose the Rolls Royce Merlin engine for the Mosquito. They had been been installed in the single-engined Hawker Hurricane and Supermarine Spitfire light fighters, were battle-tested and had proved successful in the Battle of Britain. Rolls-Royce Merlins were powerful liquid-cooled 27-litre V-12-cylinder supercharged engines, each delivering over 1,000 brake horse power in early variants, but were considered capable of development that would produce much more performance. And, of course, the Mosquito hefted two of them. Later Merlins could deliver around twice as much power as early models so the Mosquito's performance improved impressively throughout the war due to that prodigious improvement in Merlin power output.

The combination of an optimally small, light, super-streamlined airframe, was to produce the most potently versatile fighting aircraft of the Second World War. It gave the Mosquito a maximum speed of 400-475mph depending on variant. The bomber versions could carry a 4000lb bomb-load over a range of 1,500 miles. This was later increased to 8,000lbs, roughly the same as the four-engined but heavily-armed and armoured American Boeing B17 Flying Fortress; but the 'Mossie' flew much faster with much more agility than four-engined 'heavies'. The more lightly-armed fighter and fighter-bomber versions had a range of over 2,000 miles and a ceiling of about 36,000 feet. Importantly, those two engines, which gave the crews this degree of battle puissance, also gave them survivability. They could lose one engine due to mechanical failure or enemy action, and still make it back to base. Choosing an already available, successful tried and tested engine meant that de Havilland's did not have to expend valuable resources in money, skill and manpower developing one of its own. Critically, this drastically shorted time to production.

The innovative aspects of design and construction have often been described – the use of wood for instance, and the dispersed organization, but there was one aspect of the story that has received little notice. Radical innovation and widely dispersed networks of organization demand excellent coordinating management for them to work well. Spruce from Canada, balsa and other woods, suitable glues, tyres, everything from Merlin engines to Hispano cannon, from instruments to gunsights, from metal sheeting to fabric coverings for those parts of the aeroplane that could not be made from wood; and thousands of other components required for the assembly of a complex aircraft had to be sourced and procured as part of a wide supply chain. This had to be effected and supported by legal contracts that guaranteed quality, quantity, timeliness of delivery and price. It is not clear how this was done where it involved, for instance, an ex-cabinet maker working on his own with a blueprint in a garden shed; or a group of women from the Women's Institute with fabric skills. The procurement and legal departments had to be on top of the job. The assembly technicians and engineers had first to plan the assembly processes at a time before computer-aided programmes were available, logistic science was esoteric, and mathematical programme management methods such as Critical Path Analysis had not yet been developed. It is a tribute to the management skills of de Havilland's that all this was achieved. In less than two years from the first drawings, Mosquitos began to roll out of the assembly sheds.

What did the first visiting groups of observers, the men from the Ministry, the test pilots, the RAF 'Brass' (senior officers) and the Wing Commanders behold? Apart from the early marques, the two Rolls-Royce Merlin engines were placed well forward so that the pilot sat behind both propeller discs. The two three-bladed variable pitch propellers had a disc diameter of 13 feet. The fuselage was made of spruce and balsa plywood, with a top covering of fabric. Its wingspan varied slightly but in the later variants was 54 feet, the length 45 feet. The wing area was 454 square feet. It had hydraulically retractable undercarriage and brakes. The wings were manufactured not separately, but were the extremities of a single main spar of spruce, which ran right through the fuselage sitting in a slot. The fuselage was constructed of two halves divided along the longitudinal axis. When these two halves were 'slotted' onto the single main spar/wing structure they were left to sit there separated. This facilitated the installation of all the internal equipment including armament.

On completion the two halves were clamped together and secured ('glued and screwed') to make the completed fuselage, a process called 'boxing up'. A large sign in one of the assembly sheds implored,

'WHATEVER YOU DO-
DON'T SPARE THE GLUE!'

The result was a shoulder-wing cantilever monoplane with a slightly swept leading edge and a sharply tapered trailing edge. It was a strong design. The interior was somewhat cramped. In the fighter and fighter-bomber variants the crew of two entered through a door on the starboard side, and were seated side-by-side, the pilot on the left and the navigator on the right and slightly behind, protected by steel armour-plating and a flat bullet-proof windscreen. In the bomber and reconnaissance variants there was a Perspex nose-cone for bomb and camera aiming. The crew entered through a floor-hatch. The aircraft, resting on its tail wheel, appropriately seemed to gaze up at the sky. The common sentiment was that she looked deadly but beautiful too. Later, when the type had proved itself, Air Chief Marshal Sir Arthur 'Bomber' Harris, Air Officer Commanding Bomber Command, described her as '... a lethal brute with no vices.' He got the first bit right – but not so much the second, as will be described.

In summary, along with optimal size, lightness of weight, streamlining and the two Merlins, the technical advantage bestowed by the design was power-to-weight ratio. A high power to weight ratio delivered superior permutations of range, speed, operational ceiling, agility and importantly, war load.

Range meant that a wide choice of targets was available within the aircraft's maximum radius of action.

High speed meant surprise, agility, the ability to evade interception.

High operational ceilings meant imperviousness to attack by the less high-flying fighters; to be able to remain out of the effective range of anti-aircraft guns and huge radius of camera sweep for photo reconnaissance.

Hitherto unheard-of speed-altitude combinations, which were never matched by the enemy, enabled the Mosquito to dispense with heavy, drag-inducing, power demanding defensive guns and turrets, which in turn enabled two important variants – the bombers and fighter-bombers – to carry increased war loads.

Officialdom, both civilian and military was persistently sceptical, however. Some commentators have been scathing about this scepticism, but they may not have been fair in their views. Hindsight is a wonderful thing, but at the time the contracting officials had to take several factors into account including the co-ordination of the country's industrial efforts to get the optimum mix of weaponry development; and the distribution of contracts to get the quantities required; and the need to spread the risk that one or more of the aircraft manufacturers' designs might produce a dud. There was also an acute awareness that the war would bring on a massive across-the-board technical arms race to enhance performance in weaponry, nowhere more intense than in military aviation. Germany had a reputation for strong performance in these matters, so some assumed it was simply a question of time before they produced a fighter that could critically exceed the Mosquito's performance, and what price then of a bomber with no defensive guns? The Luftwaffe and German manufacturers experimented with injecting nitrous oxide into fighter engines to boost performance, but this was not sustained. It was not until developments with axial flow jet engines produced a jet fighter – the Messerschmitt 262 – that the Luftwaffe had a plane that might have outmatched the Mosquito's performance. By then it was too late to produce enough of them to have a critical affect on the outcomes of the air battles over Europe.

Accordingly, the first order, for fifty aircraft, arrived from the Air Ministry on 1 March 1940. This was a relatively small batch – the Air Ministry was indeed hedging its bets. At one point the Ministry expressed a view that de Havilland's should not expend resources on the new bomber, but concentrate on producing the aircraft already under contract, many of them training aircraft. The wording of the instruction lacked directive clarity, so it was disregarded. Sir Geoffrey, despite the question mark hanging over the project, saw the ambiguity of the situation as an opportunity. He quietly ordered his teams to carry on investing time in Mosquito construction and development.

The feared dangers manifested themselves. In the autumn of 1940 a force of Junkers 88s located the assembly factory at Hatfield and bombed it. The damage was widespread but, miraculously, none of the invaluable prototypes, of which there were three and on which most of the development work was taking place, was damaged. Work continued until the first prototype, now being carefully restored at the De Havilland Memorial Museum at Salisbury Hall near Hatfield, was readied for the first flight.

In the November of that year, Geoffrey de Havilland Junior, scion of the family and the Chief Test Pilot of the company, carried out the first test of the DH94, now officially named 'Mosquito'.

Chapter 4

The Three Prototypes

Only twenty-two months passed between the first drawing board plans being produced and the aircraft becoming a reality. This was a remarkable achievement by the design teams, the organizing management, and the often newly recruited and trained construction personnel. The de Havilland designers fortuitously found themselves ahead of the game. They had been experimenting with high levels of streamlining for wooden civil aircraft design for a number of years. As already noted, they had produced some beautifully sleek aircraft, including the twin-engined Comet, which was a largely experimental tour de force intended to break records. It did. It won the London to Melbourne race by a comfortable margin and had jaws dropping. It looked the part. As a teenager I used to ride my motorcycle past the public house opposite de Havilland's factory at Hatfield in Hertfordshire, which was dedicated to it and named, quite simply, 'The Comet'. It sported a scale model in red, just like the real ones, atop a plinth outside the pub, and I would often stop my Triumph Tiger Cub on the opposite side of the road to sit and stare at it. And, as previously mentioned, the designers and engineers had all the experience from the development of the awesomely beautiful four-engined Albatross airliner, which at one point came under serious consideration for conversion to a long-range bomber; but that had become a crowded market.

Less than a year into the project, Geoffrey de Havilland Jr. flew the maiden test flight at the controls of Mosquito Prototype 1 (Bomber). His subsequent reports were glowing, testimony to the aircraft's speed, ceiling, performance and agility. Although the original specifications focussed exclusively on the bomber variant, the de Havilland team began to develop Prototype 2 in a way that had *not* been specified – as a fighter. At the same time the bomber prototype had a few teething troubles with bomb-door and undercarriage retraction malfunctions. This was eventually made good but meanwhile another proposal emerged based on a calculation that, given the level of installed power, it should be possible to double the design bomb load from 4x 225lb bombs, and with relatively minor modifications to carry 4x 500lb bombs instead, two in the bomb-bay and two under the wings. Before Prototype 1 had even seen any operational flying, de Havilland's engineers had more than doubled its planned war load. This comprised a huge increase in potential strategic value which could not be lost, so the testing and modification programme commenced in earnest.

Meanwhile, again unbidden, others at Hatfield were considering giving the aeroplane both day and night-fighter armament. All the original vision and focus on an unarmed aircraft, while retained for the Prototype 1 (Bomber), were abandoned for Prototype 2 (Fighter). Four Browning .303 machine guns were arranged side by side on a supporting rack directly under the crew's feet. These guns had been tried and tested as the main armament of the Hawker Hurricane and the Supermarine Spitfire in the Battle of Britain. The machine guns were too lengthy for the space allocated and the muzzles of each the four barrels protruded beyond the nose, quickly nicknamed 'the moustache'.

The machine gun fires a solid round, usually made of lead, to maximize impetus and therefore disruption of any structure that it hits. It does not explode on contact – it disrupts by the kinetic energy contained by the flying round. The guns were originally designed for terrestrial service; however, those fitted to aircraft had their rate of fire increased, in this case from around 600 rounds per minute to around 900 rounds per minute. This was achieved by modifying the working parts to make them lighter, and the springs to make them stronger. The need for an increased rate of fire is that a fighter in aerial combat is pitted against a fleeting target. Once an enemy pilot knows you are about to attack he will use all the agility of his aircraft to avoid being shot at. Bringing a gunsight to bear is therefore difficult and if you do manage it you need as much lead going out at the target as possible because you probably won't get another chance. Four guns all firing together meant a total firing rate of 3,600 rounds per minute equally to 60 rounds per second and that's a lot of supersonic lead. Although the Hurricanes and Spitfires carried eight such guns to the Mosquito's four, the 'Mossie' did have one big thing going that the smaller fighters did not. The machine guns on the Hurricanes and Spitfires were buried widely out in the wings so the bullets did not strike their own single propellor. To obtain a concentration of power therefore, the guns had to be aligned slightly inwards so the rounds concentrated at an optimum range in front of the fighter to maximize hitting power at that point. This was a hit or miss affair – literally. Fire too far out and the rounds would start to diverge beyond the optimum range losing lethality; fire too close in and the concentration would not yet have been achieved. The Mosquito by contrast had all its guns aligned along the fuselage centre-line close to each other so the rounds would fly out in a concentrated straight line, full convergence already achieved. In a turning engagement the pilot still had to be skilled at deflection shooting, aiming ahead of the target so it flew into the line of fire; but the range at which the pilot could open fire and hit the target was much less critical .

That was only half the story – the fighter and the later fighter-bomber variants had cannon. Mounted in a compartment underneath the Brownings were four Hispano 20-millimetre cannon. This was a big gun, over 8 feet in length, 5 feet of which was barrel to bestow a long range. Its fire rate was 650 rounds per minute, and again, they were mounted four-square in a rack. However, the cannon bay was much larger and the whole gun was housed inside the fuselage firing out through four ports with no part of the weapon protruding. The rate of fire of these four cannon together was 2,600 rounds per minute, quite a lot less than the Brownings above, but unlike the Brownings, the Hispanos fired an explosive projectile. The 'shells' disrupted the target partly by impetus but mostly by burying themselves within the target structure at which point the internal charge would explode with a far greater lethality and destructiveness. These guns too were aligned with the aircraft's centreline with all the advantages this bestowed. The machine guns and the cannon could all be fired together.

When the flight and weapon testing of the 2nd (Fighter) Prototype was complete, the Air Ministry was finally let into the secret. They were excited by this surprise development as it might just be the answer to a very serious problem. Having lost the Battle of Britain, the German plan to invade Britain, *Operation Sea-lion*, had been shelved. Amphibious landings would turn into a rout without air supremacy, which the Luftwaffe had failed to achieve. Now, the Luftwaffe had switched from day attacks on aerodromes to night attacks on cities to try to cut their unsustainable losses in aircraft and crews. Initially, Hurricanes and Spitfires were scrambled in the hope that, despite the fact that such single-seat machines were not designed for this work, they would somehow prevail over these night raiders. They did not – single-seat fighters proved insufficiently effective, even with radar vectoring.

However, the twin-engined Bristol Beaufighter had been tried too and this looked hopeful. Twin-engined, it had a turn of speed sufficient to catch a bomber provided the ground controller was on the ball. It was more stable than a single-pilot machine both as a night-flying aircraft and a gun platform – important for night operations. It was heavily armed with four Browning 303-millimetre machine guns and four 20-millimetre Hispano cannon, similar to the configuration that was currently being incorporated into the Mosquito fighter and fighter-bomber variants, except that the machine guns were wing-mounted and so using them was range-critical.

It also carried a crew of two enabling the crew to share the high workload, and it could be fitted with the newly-developing forms of airborne-interception radar equipment (AI). The pilot was able to concentrate on accurate instrument flying under the position vectoring of the ground controller. The Ground Controller gave the pilot headings to steer ('vectors' or 'steers') and altitudes to fly. The pilot complied with these instructions by sole reference to his blind-flying instruments. The controller was attempting to turn the fighter in behind the target bomber at the same altitude or slightly below it. The navigator would be looking into the on-board radar scopes waiting for them to show the returns to take over from the ground station for the final positioning to attack. He would also give advice about throttle position, usually calling for them to be gradually closed to reduce engine power and slow the aircraft to match that of the bomber. Achieving attack range of around 400 yards, the pilot would look up for a visual contact with the bomber and, if this was achieved, would open fire. Sounds straightforward. It wasn't that easy though. It required a lot of practice.

Nevertheless, the successful engagements began to mount. It couldn't have taken Command and the Ministry long to realize that the two-crew twin-engine configuration was the answer to their prayers; and not long after to hope that whatever the Beaufighter could do, the Mosquito with its superior performance and all centreline mounted guns obviating range criticality, would be able to do even better.

Without delay, they changed the contract with de Havilland's from fifty bombers to twenty bombers and thirty night fighters instead. This new role had not originally been envisaged by the Air Ministry, but it proved a powerful, not to say opportunistic, change to the plan. It was also the beginning of the story of how a brilliant piece of aircraft design was allowed to develop into many new roles, which at times must have seemed almost without limit. In the meantime, readying the bomber variant for operations continued, but of the original fifty planes ordered, only twenty were now to be produced as bombers to thirty as night-fighter variants. The original narrowly drawn specification was rapidly broadening out to embrace wider capabilities.

There was more change to come. The attributes of this aircraft were becoming so obviously outstanding and flexible that all the previous scepticism evaporated. The government decided to stop hedging its bets and informed the company that they had decided to put in orders for de Havilland's to build large numbers at their factories in Hatfield and Canada. The Mosquito was about to arrive in substantial numbers.

What happened to Prototype Three?

Prototype Three was to remain as an unarmed aircraft carrying no defensive weaponry and developed into the photo-reconnaissance role. The contract was changed again and now was to be an order for ten bombers only, ten reconnaissance versions, and thirty night fighters. The Royal Air Force was ordered to ready itself to receive and equip squadrons once the planes had been cleared by service test pilots. The first two reconnaissance variants were delivered to the No 1 Photo Reconnaissance Unit based at RAF Benson Oxfordshire, in July 1941. The game was on.

Chapter 5

Spy Planes and Spies

Prototype 3 (Reconnaissance) was always going to be special in terms of performance. It was equipped with no machine guns, no cannon, carried no ammunition therefore, and no bombs or rocket projectiles. It carried photo-reconnaissance cameras, and they were, by comparison, hardly any weight at all. These machines, with the capacity to carry extra fuel, had a range of over 2,000 miles and a maximum ceiling in excess of 36,000 feet. The first operational Mosquito sortie was flown in September 1941, and because the final stages of development were mainly about what didn't have to be done, the reconnaissance variant beat all others to the honour. That particular mission was a reconnaissance flight to photograph enemy controlled harbours but it was just the first of a series of secret special operations flights that continued throughout the war. Its run-in to target was at a height of close to 25,000 feet; its targets were ports on the French coast and the roads and railways serving them. This was the beginning of a programme by which the RAF provided a strategically vital service to allied units both land and sea based, and of course to air units as well – the detailed photographing of enemy facilities, vessels and deployments at an unassailable speed, at great ranges and heights that gave high definition pictures embracing very wide areas. They also carried multi-million candle-power flares for night photography, so sunset did not necessarily bring the German *Wehrmacht* (Army) opportunity to avoid detection. For the first time, all the embattled areas of the European continent were being brought within regular surveillance range.

From the outset the Luftwaffe tried to attack them. The very first sortie was challenged my three Messerschmitt 109s but the 'Mossie' out-ran them as it was intended to do.The height and speed practically guaranteed that they would reach their targets – and get away and back to base in the minimum of elapsed time. Only system failure or cloud cover could frustrate such missions; however regarding the latter a few oktas of cloud around could be tactically useful to duck into if the opposition got lucky and able to challenge from an advantageous position.This did not happen often but was always a possibility, and the crews had to be ready to deal with it.The tactical savvy started to develop.

Safely back at RAF Benson the cameras with their precious film would be offloaded and carried by road to the nearby RAF Station at Medmenham just north of Marlow on the Thames.The Officers' Mess is now Danesfield House Hotel.They knew how to live those reconnaissance types. The film would be developed as quickly as possible and delivered to the teams of waiting photographic interpreters. The cameras carried by the Mosquitos were front-edge technology.They could achieve series of shots of astonishing definition.That was not all though.The cameras were designed to take not one, but two photographs of an object or scene from two slightly laterally displaced positions at precisely the same moment. When developed, these pairs of images were scrutinised through a 'Stereoscope', which looked like two microscopes mounted together on a single frame, one for each eye.Through this instrument, the two images combined into a single three-dimensional (3-D) image. The depth provided by three-dimensional imaging gave up details about size and structure that two-dimensional images lacked. Ambiguity of interpretation was dispelled so that it became possible to be sure that, for instance, an elongated object seen from 30,000 feet was definitely a rocket lying on its side rather than a torpedo, or a small barrage balloon. The discovery of the V2 rocket launching sites at Peenemunde was one of the most conspicuous successes achieved by these instruments and the deductions that they afforded. Initially dismissed as a torpedo, 3D imaging changed minds to realize that there was something being produced at this site that required the close attention of the targeting authorities of RAF Bomber Command.

Critically, the information derived could be evaluated in as close to real time as it was possible to achieve in 1941. The Mosquito was setting a new standard in a vital strategic role – near real time intelligence about where the enemy was and what they were doing. That enabled British intelligence analysts to make accurate predictions about the Germans' future intentions, which gave the armed forces an invaluable start in efforts to engage with and frustrate them.

Throughout the Second World War everything of any possible strategic and tactical value was photographed using this latest photographic camera and interpretation equipment . The definition was so intricate in fact that they defied attempts to camouflage the targets. Nothing was exempt. Night troop movements, economic targets, submarine pens, armament factories, the position of capital ships such as the *Tirpitz,* airfields, rocket launch sites, even weather systems prior to heavy bomber raids. Nothing, it seemed, was too invisible or too far away.

Developed from the last of the three original prototypes, the reconnaissance variant went into the thick of it first. While speed, height and agility delivered wonderful tactical advantages in protecting itself, this was but a means to a greater end. The most important thing was that the reconnaissance variant proved that the Mosquito could be employed as a strategic weapon and that prowess was delivered by its *range* performance. Within months of that first flight and for the first time, all the embattled parts of the European continent and the surrounding oceans and seas could be photographed in minute detail and scrutinized within a few hours.

RAF Benson's newly announced 'PRU' (Photo-Reconnaissance Unit) received two further machines designated as PR1 variants, and their crews suddenly found themselves in great demand. Their targets were initially in France, but requests increasingly shifted north eventually focussing on Norway. By the turn of the year the unit strength had grown to nine aircraft. This was still insufficient to serve urgent demand. Throughout that year four night-fighter variants and two bomber variants arrived at Benson. Some of these aircraft were not equipped for long ranges, but they were modified to make good this shortfall, and, as far as possible to fit them for reconnaissance duties.

Penetration ranges increased throughout the year with six-hour sorties over Germany and a record-breaking sortie over Italy and southern France that lasted nearly eight hours. Throughout the war the Mosquito gained a reputation for breaking records.

A PRU was formed at RAF Leuchars in Fife on the east coast of Scotland and north of Edinburgh. It was the first Mosquito unit to have squadron status bestowed, being designated Number 540 Squadron (No 540 Sqn). The squadron badge or escutcheon depicts a Mosquito – not the aircraft – but the insect! The squadron motto is '*Sine Qua Non*', which could be translated as 'Indispensable'.

Leuchars is sited in an important strategic position to the North in Scotland to concentrate on the northern North Sea. The radius of action achievable by No 530 Sqn enabled them to reconnoitre the whole of the coast of occupied Norway from Bergen down to the southern tip of Norway, and get back to Britain with the pictures. They were searching out anything significant, but especially shipping. Of the shipping targets, *Kriegsmarine* (German Navy) heavy units were of obsessive interest to the British Admiralty, which wanted to know their deployment and whenever battleships such as the *Tirpitz* and the battlecruisers and heavy cruisers looked as if they might break out of the Kattegat into the North Atlantic. The dread was no doubt enhanced by Churchill, who, as Prime Minister and a former First Lord of the Admiralty, referred to the *Tirpitz* as 'The Great Beast'. A Special Duties Squadron, No 618 Sqn, was formed with the purpose of sinking ships such as the *Tirpitz*. This squadron, in great secrecy, had its Mosquitos adapted in the same way as No 617 Sqn (The Dambusters) was having its Lancasters modified to carry bouncing bombs. No 618 Sqn's weapon weighed about half that of those used to breach the Rhur dams. However, The 'Great Beast' was eventually located by a Mosquito photo-reconnaissance spy-plane and sunk by No 617 Squadron using a heavy bomb called a Tallboy. The Tallboy anti-shipping bomb, the Lancaster-carried dam-breaching bouncing bomb, and the Mosquito-carried bouncing bomb (the Highball) were all designed by the brilliant scientist Barnes Wallace.

In 1941, while all other frenetic activity was going on concerning the development and this role of the Mosquito, at RAF Leuchars there was emerging a 'secret show'. That's what they called them then. Today we would call them special duties operations. The Station Intelligence Officer gradually became aware of rumours about a specially modified racing aircraft which could actually outpace a Spitfire. Given the DH98's provenance – the de Havilland Comet – it was about right. Then the day came when he was called out of the depths of the Ops Room because an unfamiliar looking aircraft had droned out of the cloud and landed unannounced on their airfield. He didn't know it at the time but as he walked towards the aircraft now chocked on the dispersal pan with its engines clicking and clacking as they cooled, that he was looking at a special duties Mosquito.

That particular aircraft was constructed as a fighter variant, but it had been disarmed. All four Browning machine guns and the four Hispano cannon, along with their clamping racks and ammunition containers had been removed. Had the Officer been able to see inside he would have seen no guns, but a reclining space for a passenger with a small inadequate heater and an intercom to communicate with the crew. This was, in fact, a transport aircraft but a very unusual one – it carried only one passenger. No-one would comment, but there were a few clues. These arrivals and departures tended to be at night. The passengers were generally young, tall, fit-looking men with blonde hair; and although they spoke English well, the conversation between themselves was in Norwegian or perhaps Swedish, or perhaps Danish – something Scandinavian anyway. Eventually it became clear that they were en route between German-occupied Norway and London; and then back again. These men were British secret agents and members of the Norwegian resistance. For agents to make such arduous and perilous journeys indicates that their de-briefing and re-briefing must have been of critical importance to the British intelligence agencies and extremely detailed – not the sort of exchanges that could have taken place using clandestine wireless and Morse code. Some of them were members of the Special Operations Executive (SOE). This secret force of men and women was specially trained in the techniques of espionage, sabotage, assassination, uses of codes and wireless communication, infiltration and disguise. And much more. Established on the direct instruction of Churchill it was placed under the direction of Hugh Dalton, the Minister for Economic Warfare. Their role was to infiltrate into occupied territory and watch, report back, and if ordered, to destroy key economic and military targets.

The British public had their attention romantically drawn to *Casablanca* as a nest of spies by the eponymous 1942 Warner Brothers film starring Humphrey Bogart and Ingrid Bergman. *Casablanca* was nothing compared to Stockholm as a spy capital. Unlike its brothers and sisters in Norway and Denmark, Sweden chose to remain neutral in the war that raged around them, and continued to do business with both sides in the most cataclysmic contention of the twentieth century.

Swedish neutrality had its uses to both sides. There was a continuous need for diplomatic exchange to facilitate information exchange and trade deals. Letters to prisoners of war could be sent by such channels. Coded messages posing as love letters could pass muster when passed on under the scrutiny of the salacious but suspicious. And Sweden manufactured lots of things now referred to as 'dual purpose' – items that could be used for peaceful, or if necessary warlike purposes, such as components that could fit engines to propel a fishing boat or a navy destroyer. Items like the 40mm Bofors anti-aircraft gun, purchased and used by both Axis and Allied forces and which was so effective and ahead of its time that it is still in use today. Continuous contact was necessary with such a necessary economic ally, even though it was also doing business with the enemy. Detailed contact with Russia, the mighty ally, could be safely delivered only through Stockholm. And that is to say nothing of ball bearings.

The United States Eighth Army Air Force and the RAF conducted a joint raid on Schweinfurt in1943. It was the centre for the manufacture and supply to the German armaments industry of ball bearings. The argument was that if you destroyed the enemy's ability to produce ball bearings, everything would literally grind to a halt. There was condemnation of this argument by Air Chief Marshal Sir Arthur 'Bomber' Harris who called it 'a typical panacea target', arguing to the contrary that the Germans must have thought of this possibility and made preparation to get around it. He did not prevail, the bombing went ahead; but the British Foreign office set up a secret operation to make a deal with the Swedes. In June 1943 two Mosquitos took off from Scotland, and with two senior civil servants shivering in the bomb-bays, after what seemed a very long flight, landed in Sweden. They negotiated this monopsonic agreement with the Swedish government to buy every ball bearing that Sweden produced. Germany would get none, which was the real point of the transaction. The deal was contracted only a few hours before German representatives, after a much shorter and less dangerous flight, arrived. They returned to the Reich empty-handed.

The British Overseas Airways Corporation (BOAC) was tasked with flying daily flights into Sweden. The flights were dangerous with inadequately performing lumbering Hudsons. The BOAC pilots had no confidence, as they penetrated the Skagerrak, that their assumed neutrality would be respected if challenged by a Fw190 'Long Nose'. But what about this new thing? This DH Mosquito plane? BOAC did an assessment and said 'We want it!' Refused initially, persistence paid off and they got it. BOAC pilots continued this service with Mosquitos to the end of the war. They flew 520 round trips in just over two years. Four aircraft went missing with the loss of eight crew and two passengers. These were special duties operations and detailed information about the cargoes they carried is, unsurprisingly, scant. They were brave men.

Chapter 6

Fighting in the Dark

It is hard to imagine what might have been the British Government's plan to defend the towns, cities, docks and industrial complexes of Britain against air attacks delivered by an enemy from the air at night, assuming there was a plan at all. It looks more like trial and error in some respects. The fighter programmes had all been specified to produce day-fighters. It was assumed that most bomber attacks would be during daylight, and certainly that had been borne out by the pattern of raiding so far. But the RAF day-fighter planes and the pilots who flew them under ground radar guidance had, after a gruelling struggle, proved fit for purpose. In The Battle of Britain, Great Britain had been saved from Nazi enslavement by men and women often still in their teens. The Luftwaffe, surprised, switched to seeking the cover of attacking at night to cut the number of aircraft shot down and the number of crews killed or captured.

The successes of the planning, organization and operation of the daytime defensive systems were not matched by arrangements for night defence. The Luftwaffe, however, along with its industrial and scientific support base, had anticipated that accurate night bombing would be essential whatever the theatre and therefore transcended tactics – it was a strategic issue. So they went to work to deliver the capability.

Changing to night operations initially brought the Luftwaffe successes that alarmed the British. There must have been feverish discussion and theorising, but whatever possibilities were being envisaged, it took some time for the British to develop the ability to meet the threat at night.

The initial stop-gap conclusion was to try to teach day-fighter pilots to fly in the dark. One of the first such attempts was with day-fighters such as the Hawker Hurricane, the Supermarine Spitfire, and an extraordinary-looking hybrid called the Boulton-Paul Defiant. This latter looked superficially like a Hurricane but with a rotating gun turret embedded into the upper fuselage behind the cockpit to enable the gunner to open fire around an enlarged arc. Full marks for innovation, but this did not work out well in day fighting because it was vulnerable to the superior performance of the Luftwaffe's Messerschmitt 109. This German aircraft, the top fighter machine of the Luftwaffe at the time, had forward firing wing mounted guns and a single centre-mounted cannon tucked into the upside down Bee engine and firing through the centre of the propellor boss. The Defiant had neither. It fared better when converted to night fighting, but could not by any means be regarded as an answer to the threat.

The Hurricane and Spitfire also tried .The main problem remained that human sight has limitations. It needs light to illuminate objects, which are reflected into the eyeball to form an image on the retina, and then a brain to make sense of it. The light might come from the sun, or the moon, or a torch, or a car headlight, or a fire or an electric light – but there has to be light; and compared with many other animals, humans need a lot of it. Within a comparative visual spectrum of mammals we are low performers. Outside the visual spectrum we are almost totally reliant on sound, and we are not good at that either, especially when sitting between two roaring piston engines and propellors. Some animals such as rattlesnakes and pit vipers have evolved a natural thermal-imaging ability enabling them to detect prey from yards away in total darkness using only the infra-red bands within the electromagnetic spectrum. We can't. Owls and lemurs, with their huge eyes and dilating pupils, can operate in low levels of light that to us are abysmal. Bats can fly in total darkness, avoiding collision because they can use seven octaves of squeaking as a form of natural airborne sonar. Airborne sonar. Flying in darkness. It may have been that bats provided a clue as to how successful air night fighting might be achieved.

Much later in the 1970s experimental night vision goggles began to become available that used enhanced light from sources such as the stars or masked torches. They also had an infra-red capability. Special Duties aircrew such as myself participated in the development of both the equipment and the techniques to use them. Today they comprise an important item of standard equipment. In the 1940s however, no such apparatus was available. Operating at night was reliant on what RAF crews referred to as 'Mark One Eyeball'. The moon was obviously important, but it was a mixed blessing. On forecast clear nights operations might best be conducted during the waxing and waning half-Moon phases. Too much moonlight posed a risk that the intruding aircraft would be seen and engaged by a fighter or anti-aircraft artillery. Too little and the ability of the Intruder to find the target was difficult. A new moon in particular made navigation next to impossible.

There were other problems. Single-engined fighters were relatively unstable. In the 1930s, if you were designing a plane that had to be the acme of manoeuvrability to do its job, by design it had to have limited amounts of stability around the turning axes – rolling left and right, and pitching up and down. This makes it twitchy and more difficult to fly accurately when blind-flying. Put it on a heading, or an attitude, and it will wander off even in calm air, and air is hardly ever calm. The flight path has to be monitored constantly and deviations corrected without reference to visual outside cues; only to 'the instruments'.

These instruments, in RAF aircraft, were clustered together in the same arrangement, known as the 'Instrument, or 'Blind-flying' panel. At the centre top was the Artificial Horizon (AH), sometimes referred to as the 'Master Instrument'. To its right was the Rate of Climb and Descent Indicator (RCDI). To its left was the Air Speed Indicator (ASI) calibrated in statutory miles per hour, not knots – nautical miles per hour as in modern aircraft. Below these three was another row and on the left was the Altimeter, which was an instrument responsive to air pressure, but which could interpret this in terms of height above ground level or altitude above sea-level depending on the setting of a sub-scale, adjustable by a twisting knob. In the centre of the bottom row was a Direction Indicator (DI) to show the aircraft's heading. This was a directional gyroscope which would maintain its orientation while the aircraft turned around it. The presentation was clearly marked in degrees of the compass, but aircrew paid a price for this clarity caused by 'gyroscopic precession'. While the gyroscope exhibits a mysterious entanglement with impossibly distant stars enabling it to maintain its orientation with relation to them, the Earth does not. The Earth spins on its axis at a rotational velocity of 15 degrees every hour, which causes a gyro-compass to apparently 'wander off' the selected datum by a degree every four minutes. This was persistently inconvenient when trying to navigate, because the DI had to be continually realigned with the main magnetic compass, which referenced the earth's magnetic field. This main compass was also sometimes mounted somewhere between the pilot's legs, which is not ideal. Putting your head down and peering between your legs while flying a fighter aircraft could induce disorientation because it could confuse the balancing mechanisms in the inner ear; and besides, it was terribly undignified. However, the Earth's rotation does confer night and day, and as this was generally regarded as a good thing, one imagines few complained.

Finally, there was a 'Turn and Slip Indicator' (TSI), a precise instrument that showed the pilot when he was not flying straight and/or balanced in yaw. Its most valuable contribution was made if you seriously mishandled or were blown by flack into a spin. Without labouring the point, a spin at night or in cloud means death very shortly if you don't deal with it, and you can only do that with reference to the TSI. Apart from spin recoveries, yaw is what you do when you go around a corner in a car. In an aircraft en route its presence is undesirable for a number of reasons including because you might not be pointing in the right direction to maintain course even though the compass says you are. So night/blind flying can be tricky stuff taken all together, requiring high levels of training and high and prolonged levels of concentration, especially when you are also having to monitor the temperature and pressures within your engine(s) and any critical gearboxes, electric amperages, pneumatic and hydraulic pressures, and critically your fuel state and whether or not, given the winds and weather encountered, you have enough to continue to try to do what you were told to do. Turbulence aggravates matters, as does engine and airframe icing. The list is not exhaustive, but it shows that pilot workload certainly could be exhausting and in extremis could overcome the pilot's ability to cope. For night-fighting, a 'platform' would have to be found or developed that was more stable than a single-engined day fighter and which could be flown with much more manageable levels of pilot workload.

While Sir Geoffrey de Havilland was applying his innovative brilliance to the design and production of his 'Wooden Wonder', Bristol Aircraft were engaged in doing something that was also remarkably clever. They had been asked to consider how to address the issue of air-fighting during the hours of darkness. Their first response was a modification programme to convert the Bristol Blenheim bomber into a night fighter. Trials were carried out with some success. The Blenheim was a considerably bigger aircraft than a Hawker Hurricane, with a bigger wingspan and control surfaces. It was also heavier and the greater inertia added more built-in stability making it easier to control under blind-flying conditions. It was also twin-engined providing two lines of propellor thrust either side of the fuselage, which, when synchronised, added further to stability.

Importantly, it had the internal capacity to carry and experiment with some new secret equipment called by various names, but eventually to be dubbed Airborne Radar (AI). It served as a test-bed and concept prover, and did succeed in shooting down some bombers at night, but it had limitations in the role. In particular. the increased performance of Luftwaffe medium bombers such as the Heinkel 1-11, and the Junkers 88, meant that a faster platform would have to be found.

Bristol Aircraft selected their torpedo-bomber, the Beaufort, for modification to fill this role. The Germans did something similar with the Junkers 88, which began life as a medium bomber and was modified to produce a successful night fighter. Role-modification of an existing aircraft met the need for a quick, highly cost-effective answer to an urgent problem. This first Bristol night-fighter variant was introduced in 1940. They named it, as noted,the 'Beaufighter'. The torpedo-bomber had been stripped down and drastically lightened. Initially it retained a mid-upper defensive gun turret, but later these were often removed adding to the speed, rate of climb, range and endurance of the aircraft. The pilot sat in a single seat in the middle-front of the aircraft. These aircraft were armed with Browning .303mm machine guns embedded in the wings and firing outside the propellor discs. This meant that the problem of range/convergence, already described, applied when using these guns. However, in exactly the same format later used in the Mosquito, four Hispano 20mm cannon were enclosed in the belly. These were centre-line aligned and so entailed no range/convergence issues. Thoughts were on night-fighting initially, less on the fact that these variants would also comprise effective ground-attack aircraft. It was estimated that the fire-power of a Beaufighter with all forward guns firing was equivalent to a broadside from a light cruiser. A somewhat whimsical comparison perhaps, but it did make a point, no doubt duly noted by de Havilland's, who were now hard on Bristol's heels with their own brainchild, now being called the 'Mosquito'.

Bristol also avoided the cost and delays of designing new engines, but unlike de Havilland's who bought Merlins from Rolls-Royce, they selected a successful aero-engine that they themselves had designed and were already manufacturing – the Bristol Hercules. These were 14-cylinder, 24-litre, double-banked, air-cooled, supercharged, radial engines. They were used to power many successful types including the Vickers Wellington, the Short Stirling, the

Handley Page Halifax, and the B11 Avro Lancaster. They were reliable, powerful, and lacking the need for a coolant jacket required by liquid-cooled engines such as the Merlin, could absorb impressive amounts of battle damage and still keep going.

So, the beginning of effective night-fighting by RAF Squadrons in the Second World War was to start with the Bristol Beaufighter. The 'Beau' led the way. When the new Mosquitos started to arrive at the squadrons offering an even better performance than the Beaufighter it is unsurprising that many of the men chosen to fly them had gained valuable experience from being in on the operational development of night fighting techniques and tactics on the Beaufighter. The legendary John Cunningham was one of these pilots. Flight Lieutenant (later Group Captain) Cunningham, with his crewman Jimmy Rawnsley, were emerging as the top-scoring night-fighter crew of these battles. The press and the public information machine became aware and concocted a daft story that they had an addictive appetite for carrots, which bestowed incredible night vision. There was some scientific evidence for that based on the chemical composition of the eponymous carotene found in carrots – but not much. The British public however, chose to believe it; the Germans probably enjoyed a good laugh and the man himself gained the sobriquet of 'Cat's Eyes Cunningham'. He hated it. Cunningham, put into a firing position by Jimmy Rawnsley, tore apart whatever was in front of him with cannon fire. Fly back to base, land, bacon and eggs, go to bed and sleep. Same again tomorrow.

Chapter 7

From the Ground Up

Luftwaffe High Command, in changing from day to night bombing raids, encountered the problems that were to face RAF Bomber Command when it did the same thing later in the war and for the same reasons. Bomber Command suffered from the disadvantages that, initially, it had available no battle-tested operational navigation equipment enabling it to acquire and therefore attack the targets; nothing sophisticated enough to enable accurate blind bombing and no way of confirming objectively what had been hit. Crew reports of their performance were treated with increasing scepticism. *The Butt Report*, published in August 1941, revealed the overall failure of RAF bombers to hit their targets. This Report estimated that of the aircraft claiming to have made a successful attack on its target, only one in three actually got within five miles of it. Fitting downward pointing cameras, triggered by bomb release, provided supporting evidence of bombing inaccuracy.

The Luftwaffe's change of tactics was backed up by some realistic thinking. London remained an easy target because of its size, because it was on the coast and had a big river running through it, visible in most moon-phases. Finding smaller inland towns or industrial complexes at longer ranges and with no features visible at night was a more difficult challenge. However, German scientists knew that it was possible for transmissions, or 'broadcasts' to indicate direction.

Suitable receivers at a distance were able to detect the direction of the source of the broadcast, and therefore its relative bearing to, in this case, an aircraft. Marrying this information to compass bearings enabled an aircraft to fly outbound on a specific track over the ground by correcting for any drift off the track required caused by wind change or, dare one mention it, by less than skilful instrument flying. Range could be determined by using other suitably placed ground transmitters so that their transmissions, with the outbound 'beam', formed a grid or lattice. Locked onto the radio-determined outbound beam, the range was determined by when the correct cross-bearing was reached. When these two pieces of directional information coincided, this comprised an accurate 'fix' of position. The fix could be adjusted from the ground to give an accurate position for bomb release based on trajectory calculations taking into account height of the aircraft, speed, wind at height, weight of the weapon, weapon flight characteristics etc. German scientists and technicians honed it into an impressively effective system, as the British discovered to their dismay when it was used in attacks such as those on Coventry in November 1940.

German scientists began development of navigation and target location by transmitted radio paths known as 'beams'. They achieved promising success in one of their first programmes code-named 'Knickerbein' (Crooked Leg). This was developed in a number of exploratory directions. The best of these evolved into a very accurate bombing system, code-named 'X-GERÄT'. It wasn't at all a good code-word if by 'code' you were seeking a word that would neither indicate what it was nor reveal its purpose. 'Gerät' means 'apparatus' in German, and it wouldn't take much guess-work to suggest the possibility that the 'X' could be something to do with cross-bearings, a common navigational technique for fixing position. By transmitting a grid of crossing beams an accurate track over the ground and an accurate release point to hit the target could be achieved. Night bombing accuracies began to rival or even exceed those of day attacks, but in order to promote even better results the Luftwaffe used one of its very best air groups, *Kampfgruppe* 100(KGr100). This was the first use of what the RAF, much later, would call a 'Pathfinder force'. K.Gr 100, well practised in target location using this radio system, found the targets and marked them for the following main bomber force by using incendiaries and flares. The following main force then simply had to check that the ground fires were where *X-Gerat* said they should be, and bomb using the fires or flares as target markers.

It was an effective system and it had to be disrupted. The British poured a lot of resources into detecting the radio transmission systems, and then interfering with them. Eventually, at the end of 1940 a Heinkel 111 bomber with *X-Gerat* equipment crashed in Britain and the equipment was removed and analysed. The analysis was used to construct radios for interference with the cross-track signals indicating range to target. The equipment could be fooled into believing that its range outbound was greater than was actually the case and, as the bomb release mechanism was automatic, the bombs were released too early inducing an inaccuracy of about a kilometre from the target. The Luftwaffe, having quarrelled amongst itself about whether their system was unreliable or whether the British were tampering with it, were eventually driven to conclude (correctly) that the latter might be the case and thereby found itself engaged in developing counter-measures to further improvements to British jamming techniques. This became known as the 'Battle of the Beams'.

A 1977 BBC series entitled *The Secret War* documented these secret technical developments achieved by both the German scientists and their British counterparts. In 1977, many of the participants were still alive and definitely kicking, including the star of the show Professor R.V. Jones of Aberdeen University, so it did not lack punch or authenticity. The deception shenanigans to counteract *X-Gerat* under the direction of Professor Jones were described for the first time vividly to millions on the small screen. It contains such gems as when Jones, having discovered that the powerful BBC TV transmitter at Alexandra Palace in north London was lying completely idle, with an ease that surprised everyone including himself, co-opted it for use testing and operationalising his jamming programmes. He had a sense of humour and thought of his programme as legalised mischief.

Professor Jones started this work investigating and discovering how to jam the predecessor system to *X-Gerat* – code name '*Knickerbein*' (Crooked Leg). The countermeasure work on jamming *Knickerbein* was, of course, top secret, hidden from everyone without a strict need to know – there were to be absolutely no exceptions. Except that apparently there was at least one exception. Professor Jones reported that he and his staff received a visit from Winston Churchill. The Prime Minister walked into the room, removed his cigar, winked at Jones, and growled 'Hello Professor – how's your knickers?'

At the same time as all this was happening in the back rooms, it was also realized that another key objective in the night skies would be to try to shoot down the Luftwaffe's crack "path-finders" – the crews of the KGr100. They were making a material contribution to the accuracy of Luftwaffe night bombing. Shooting down these-target marking experts would significantly degrade the bombing accuracy of the main attacking force. Hunting the Luftwaffe pathfinder crews rose to the top of the agenda. As the practice and experience grew, it evolved into a slick operation.

* * *

What follows in this chapter is a typical account of operations -'Ops' - based on the author's own flying experience. And see also similar sections in chapters 8, 9 and 12.

The pilot and navigator clumped out across a muddy grass area towards a dispersal pan in the fading light towards the night-fighter. They felt the weight of their parachutes thrown over their shoulders, the sucking mud beneath their heavy flying boots, and their navigation bags full of maps and frequency documents. They paused briefly to chat with the Sergeant who was responsible for their departure. In the warmth of the Operations hut they had studied the briefing containing meteorological information and little else. The cloud base was at 2000 feet, the temperature plus 8 (degrees Centigrade); the visibility 2000 yards; the wind light and variable. The pilot walked around the Mosquito doing the external check of tyres, no fuel leaks, everything secure and so on. The navigator was already in the aircraft, Nav-bag open, putting documents on his little desk. The recently refuelled aircraft had a whiff of petrol about it. Gingerly selecting the Battery Master Switch to ON, the aircraft internal lights illuminated and the aircraft's electrical systems came alive, enabling him to start the internal checks of his equipment. The radios crackled and shared their half-finished exchanges.

The gyroscopes started whirring up from rest, causing the blind flying instruments to wake up and eventually stabilise. The pilot threw his equipment up through the hatch and climbed in after it. He heaved himself into the pilot's seat and strapped himself in. He pulled the radio frequency card from his case and set the first frequencies. He broke the silence with the Before Start Checks, which had to be recited verbally so that the navigator could hear and respond as appropriate. From that moment there were no individuals any more. Instead, a single interacting crew intent on a result. The navigator continued listening out on the radio ground frequency while checking his equipment. The pilot was at his busiest time with aircraft flight preparation, and within three minutes had performed over fifty checks and tests. The Sergeant looked on intently, occasionally glancing up at the gathering cloud and rapidly darkening sky.

'Ready to start?'
'Ready.'

The pilot primed the port engine with a good sluice of fuel. He looked at the Sergeant standing with the fire extinguisher in front of the Rolls-Royce Merlin 25 engine on his port wing. He pointed at the engine. The Sergeant held up a thumb, crouched down. Check: Fuel Cock ON; Fuel Pumps ON. He pressed the START button. The engine whined like a big metal dog, coughed, misfired in a cloud of blue smoke while the three blades of the propellor jerked and slowed, jerked and slowed, until it all went right. The engine roared and the propellor suddenly became invisible. The aircraft began to vibrate and judder. The sound was enormous. The pilot carried out the after start checks; oil pressure, fuel pressure, RPM, electrics all on line, all warning lights extinguished, hydraulics, pneumatics…

He pointed at the starboard engine. The Sergeant walked the fire extinguisher around the front of the aircraft to the starboard engine, held up a thumb, crouched down. Check Fuel Cock ON; Fuel Pumps ON. He pressed the START button. The engine whined, coughed, misfired in more blue smoke while the propellor jerked and slowed as seemed traditional, until it all went right. The propellor became invisible as the second engine roared, and the aircraft juddering increased. The sound was now totally deafening to anything that was not coming into his leather flying helmet through the intercom or from the radios. He carried out the after start checks; oil pressure, fuel pressure, RPM…and eventually, all complete.

'Ready to go?'

'Yes ready to go. They want us to the south-east initially to 8000 feet to see if we can catch any early starters against the setting sun. They'll call us if they have trade.'

'Trade' was RAF slang for hostile radar blips. The daylight was fading. The ground-crew pulled the chocks away and ran aside. The pilot gave an insouciant wave, released the brakes, eased forward the throttles…and Mosquito Q for Queen, slowly, started to move.

During the ground taxi prior to the line-up for take-off, the pilot completed his take-off checks, which included engines run-up while holding position with his brakes, check of the dual magnetos, and a long string of others. These included the all-important instrument checks. Was the Artificial Horizon indicating nose up and wings level? Was the Direction Indicator aligned with the compass heading? Did the Altimeter have the correct sub-scale setting, and was it reading zero so after take-off it would show the height above the aerodrome in case an emergency necessitating flying a circuit for an immediate re-landing back at base? Much more. These checks of the instrument flying panel were vital (literally referred to as 'Vital Actions') because as soon as the aircraft became airborne, all those useful little lights on the runway edges would quickly vanish below and the aircraft would be enveloped in an inky darkness.

The pilot positioned the Mosquito at the take-off point lined up on the grass runway heading. The Air Traffic Control tower shone a steady green Aldis lamp at him giving him clearance to take-off. He held her on the brakes and smoothly, slowly opened the throttles. The Mosquito propellors both turned in a clockwise direction inducing a torque swing to the left as the power increased. To counteract this the pilot twisted his left throttle hand so deliberately applying more power to the left engine; his right leg was also applying full right rudder so as the rudder bit the air with increasing speed it would also correct to the right. Brakes off. She started a forward rush and for once kept straight in the first fifteen seconds. The wind must be a bit from the right. The rudder became effective enabling full power now on both engines. He lifted her tail wheel, paused, and then started to ease back on the control column to unstick her. The bumping over the ground suddenly ceased. She was airborne.

The pilot's eyes locked onto the instruments. A quick glance at the RCDI to confirm that they were indeed gaining height and the pilot raised the undercarriage while immediately starting a methodical scan of his instruments. It was a technique known as the 'radial scan' because it centred on the Artificial Horizon which showed the all-important aircraft attitude. Get the attitude right for any phase of flight – climbing, level flight, descending, turning – and things settled down as a consequence. Get it wrong and you created all sorts of equally consequential problems that had to be eradicated before you could claim that you were properly in control of the aircraft. The pilot knew the aircraft was just a platform to enable the achievement of other things. Navigators needed the pilot to settle things down to reduce the variables in calculating an adjustment to maintain, or regain, the required track and an amended estimated Time Over Target (TOT). The bomb aimer needed a settled-down aircraft to enable the bomb-sight adjustments to stay valid long enough to score hits on targets. The night-fighter navigator air-interception operator wanted no surprises in the run up to target. The pilot was called Michael Sykes. He had been educated at a grammar school near Kingston-on-Thames. 23 years old, he had recently been promoted Flight Lieutenant. He preferred to be called 'Mike'.

The navigator was called James Montague. He was educated at a public school near Rugby. He was 21 years old, a Flying Officer. He preferred to be called 'James'.

Notwithstanding his juniority, the navigator was really the star of the show. He was responsible for the navigation of the aircraft, and for communications often including wireless telegraphy requiring the practised skills of Morse Code. He also had to re-load the four cannon in an air battle when their ammunition was spent and when the fighter was cavorting all over the sky pulling high 'g-forces' that made the cannon magazines seem like they weighed a ton. He had to dispense raisins from small paper bags, chocolate and tea. He was also required to be an air interception radar controller – of his own aircraft. The ground controller's job was to calculate the intended track of the incoming bombers and thereafter vector the fighter into position for the all important closing turn, during which the navigator would be instructed to 'Flash your weapon' (turn on your

Airborne Radar). As the fighter rolled out of the closing turn the ground controller would tell them that the target aircraft was in front and indicate the range and direction.

There had been excellent improvements in the ground radar's performance over the past year. The ground controller could now do much more than putting them in the 'ballpark' – he could put them on the ball. He could position them in the intercepting turn and give an accurate roll out at a point where the chances of the airborne interception radar to pick up the target were greatly enhanced. The whole operation had gradually transformed itself to a point where the crews became confident they would get a flawless handover to an attack. The stressful mismatches of the early days had caused frustration and anxiety, not to mention self-doubt, to the two-man crews who were doing their absolute best while still having to cope with the repeated disappointments of unsuccessful interceptions.

A successful attack occurred when the ground controller gave you a good turn on to the bomber's heading that put you less than a mile behind it. It happened when, as the pilot straightened out from the turn, the navigator, eyes pressed to the eye pieces, saw an azimuthal blip appear on the scope moving slowly down the centre-line indicating a slowly closing range between the bomber and the stalking night fighter. It happened when the other scope indicated that the altitudes of both aircraft were roughly matched.

The navigator, all thoughts of navigation, raisins and tea forgotten, had transformed himself into an airborne interception radar operator. He peered into the two scopes of his radar set. The target was within azimuthal detection and within a few seconds showed up as a blip distinct from all the interference and random returns that had plagued the early developmental equipment. The navigator began to talk the pilot through changes of heading, left or right, to bring the blip sliding slowly down the scope centreline. The other scope indicated the target's altitude relative to the Mosquito. The navigator had to appreciate rapidly the inter-relationship between these two data, and issue instructions accordingly. The target was low and the fighter was still smartly closing the range. In this situation you could not simply dive off the height – even with closed throttles the aircraft still accelerated sufficiently for there to be risk of overshooting the target.

'Throttle back Mike – lots please. Dump height.'
'Roger – back to 2500 rpm.'
'Make it 1500.'
'Ok – they won't like it.'

James thought, 'Won't like it? They'll have to lump it. What do we think this is – a bloody air-test?'

'They' were two Rolls-Royce Merlin 25s that sat outside them on the wooden wings. They were 27 litre twelve cylinder piston aero engines, the cylinders arranged in an inverted 'V'. They were liquid cooled and supercharged, and they did not like low RPMs. The twenty-four spark plugs started to oil up and the engines put in the odd misfire, which made the pilot wince. Each backfire could cause a tell-tale flash of flame that could and just might draw the attention of the bomber rear gunner, now less than 700 yards away. The reduction in power was to reduce the rate of closure while at the same time losing height. The Mosquito had no airbrakes and sometimes resorted to 'dumping' (extending) the undercarriage as a substitute to create more drag for a greater rate of sink while reducing the rate of closure. That could become tense until it stabilised, but if the Mossie's speed and height were too great to make an attack it would have to break away to avoid a collision or blowing past the target bomber right into the arc of fire of the bomber's front guns. Breaking off meant you had lost that bomber for good. By the time you had turned back to try to re-engage it again it could be miles away. So these closing calls by the navigator were critical both in terms of direction and particularly in terms of timing.

James said,

> 'Good heading. We are still too high but rate of closure alright if you can get rid of that height. Lose more height please.'

Mike extended his left hand and selected undercarriage to DOWN. Maintaining the speed, he saw RCDI needle creep down to indicate a descent which reached 500 feet per Minute.

> 'That's nice. Coming level with him now.'

There was a sudden yawing movement and momentary change of note from the propellors. This confirmed that the descending fighter had just dropped into the bomber's slip-stream – a more or less perfect line up. The bomber should now be at 12 o'clock going slightly high at 500 yards and closing.

> 'Level off please'.
> 'Levelling off.'

Mike eased up the throttles. The Merlins stopped coughing and complaining as they were pushed back up towards the cruise range. The pilot was still glued to the instruments as tightly as the navigator was to the A.I radar set. Nobody was looking outside the aircraft into the blackness as the Mosquito manoeuvred in a *danse macabre* behind the bomber.

At about 400 yards James took a deep breath and said 'Look up – you should be able to see him.'

The pilot, still glued to the blind-flying instruments, flicked his eyes upward without moving his head. And there, now less than a quarter of a mile in front were the glowing exhausts of two engines. He made slight adjustments in yaw with his rudders and a slight roll and back again to move the line of his aircraft to exactly correct any lateral drift on the target.

These final sighting adjustments were moments of intense danger. If the bomber's rear gunner spotted the stalking night-fighter's exhausts he would open fire at little more than point blank range. But if he did not, if the crew were eating raisins out of a paper bag and drinking ersatz coffee, one can only hope for them that they tasted good.

The pilot had already turned the Gun Selector to FIRE. Raising the Mosquito's nose an almost imperceptible amount, he tensed up as he placed his gloved forefinger around the cannon firing trigger and watched the Heinkel 1-11 sink slowly into his gun-sight.

Chapter 8

Deadly Gremlins*

A few pilots stood out from the others and there was a tendency on the part of squadron commanders to place the responsibility for the required flight testing of aircraft delivered to their squadrons on them. Aircraft produced at factories in places such as Bristol and Hatfield had to be dispersed away as soon as their construction was completed to prevent them being destroyed by Luftwaffe attacks. The pressure to get them out was relentless. They arrived brand new from the manufacturers often flown carefully and gently by the women of the Air Transport Auxiliary. These flights were high risk, and accidents occurred in the process of delivery of new aircraft to the squadron bases. This was simply because the highly complex manufacturing processes that the production of aircraft required, especially under the pressure of war, entailed a probability that something critical could be missed. If the delivery flight passed without incident, another risky bit could begin. This would comprise flight testing the aircraft to assess its fitness for battle and that involved stressing the aircraft close to its limits. If there was a fault lurking, the delivery pilot or the test pilot would find it and that might cost him, or her, injury or worse.

* Gremlin – a supernatural being that malignantly interfered with aircraft systems. They had various names; for instance the one that mysteriously increased an engine's fuel consumption was called 'The Petrol Boozer.'

To mitigate these hazards a test-pilot school was set up at Boscombe Down in Wiltshire to provide specialist training to test pilots, but that did not happen until 1943. In the years before then you could be rewarded for being regarded as the best pilot on the squadron by a lot of test flying. These pilots did get due recognition eventually after the war. Adults in the 1950s still remembered names like John Cunningham, Neville Duke, John Derry, Peter Twiss, Brian Trubshaw, Roly Beamont, and Eric 'Winkle' Brown. They were the rock stars of their day at the zenith of British aircraft innovation. 'Winkle' Brown, who was known to me, goes down in history for so many things, but must be mentioned here because he was the first pilot to land and take off a 'navalised' Mosquito from an aircraft carrier. He would not have appreciated being compared to a rock star.

* * *

The instructor was a former Beaufighter pilot who had been caught by a German flak cruiser and lost part of his face. Despite his injuries he was returned to duty and he was grateful. He had been noted as a gifted pilot and got more than his fair share of test flying. He was now an instructor. He was still fascinated by aeroplanes, but his time engaged in flight testing had instilled caution. Instead of the sense of awe that he had felt as he first approached a Hawker Hurricane, he now felt a highly assertive wariness. The student, a Flight Lieutenant, walked in step with him over the grass airfield towards the Mosquito sitting silently waiting.

'How did you like Beaufighters?' the instructor asked.

'Loved them – I almost felt safe.'

'Do you hunt?'

'Er – yes, I did a bit with the Bicester and Whaddon Chase.'

'Ever catch anything?'

'Fox you mean?-no never did. Didn't care actually. I was more interested in dancing with the Hunt girls. They do have some lovely looking women in Stratton Audley.'

'Well listen to this. Your training course is going to be short because you almost completed a tour on Beaufighters. You will be familiar with night-fighter tactics so we won't need to tell you how to suck eggs. Let me put the comparison between those two aircraft this way. You have ridden hunters. Now you are going to ride a race horse. I am going to put more power in your hands that you will know what to do with until you master it. Chuck your gear up and get into the cockpit. Remember that this plane has killed several experienced pilots because they weren't aware of what to expect. Strap and we'll talk'.

They strapped in.

'Yes it's brilliant alright, but a word of warning. There is a view growing that this aircraft is perfect. De Havilland's are calling it "All things to all men." In that respect it reminds me of the Spit. Ever flown a Spit?'

'No.'

'That aircraft has an adoring fan club. People seem to want to believe in her. The trouble with fascination is that it makes you overlook the bad points and then you can get badly caught out. I'm an ex-test pilot and I have had all the sentimentality about aircraft frightened out of me. There is no room for sentimentality in this business. That includes this aircraft. OK?'

'OK.'

'She affords you single engined safety – the ability to get an engine shot out and fly back to Blighty on one engine. Here's the trap; when arrive back on one you have managed 90 per cent of the emergency. It's the other 10 per cent that could kill you. And your navigator. That will be a tragic own goal. One up to the Luftwaffe and they never fired a shot! Think about it.'

There was a moment of reflection.

'We will be spending a few hours training and a lot of it will be stooging about on a single engine. There is one thing I need to embed in your brain – if you are down to one engine and on final approach, for God's sake don't get low and don't get slow. You need to be on the hot and high side with the good engine at a lowish throttle setting. It's almost a glide approach. Keep her high enough on the final round-out so you don't need any more than the merest tickle of throttle and best of all not at all. Plenty of height in hand will also deal with any down-draft on finals. Once you have set her up with a throttle setting on a single engine finals aim to land without moving that throttle forward again. She will try to roll over and that's not how you want to be when descending through 200 feet. Got it?'

'Got it.'

'It's all a question of torque. On one engine you are flying asymmetrically with the good engine trying to yaw you away from itself. If the good engine is to port it will yaw starboard and vice versa. In flight you can correct this by applying rudder towards the good engine. However, as you lose speed, as for instance in an approach to landing, the rudder loses its effectiveness and even with full rudder applied you will still start to crab sideways. You will be tempted to correct that yaw with roll. You might get away with that that technique with other aircraft, but you won't get away with it in a Mosquito. If you feel that loss of yaw control when you are making an approach, do NOT correct with aileron. Put the nose further down to get more airspeed and reduce throttle. It is better to shut the good engine down and do a glide in or even crash landing straight ahead than try to power your way out of it. Odds are you will not make it'.

'Well,' said the student, 'I thought Butch* said she had no vices.'

'When you have that amount of brass on your hat it only guarantees that your opinions won't be contradicted. Unfortunately for the rest of us, it doesn't guarantee those opinions will be right – this being a case in point. Sorry to be so repetitive but believe me Chum – you've never met anyone that has your interests at heart more than me. Not even your Mother. Get the single-engine approach procedure etched in your brain. Enough chat – let's get going!

The student started to work his way through the check-list. Taken directly from the Pilots' Notes Manual, it was printed on sheets of paper that were secured in a cardboard folder. The Instructor had advised that the normal operations checks should be done from the folder, but the emergency checks must be committed to memory. If something went wrong there would be no time for fumbling your way through books.

Checks complete, he looked at the sergeant standing outside his port-side engine, and put up his thumb. The sergeant pointed to the port engine and made a circling movement with his finger – clear to start the engine.

Ten minutes later with both engines running, he put the thumbs-out sign for 'Chocks away!' The chocks were pulled out and put in

* 'Butch' – Air Chief Marshal Sir Arthur 'Bomber' Harris, Air Officer Commanding Royal Air Force Bomber Command. His crews affectionally called him 'Butch'.

a place where the pilot could see them. No room for error. If you applied power with the chocks in you could nose the aircraft over and then all hell breaks loose.

The ATC tower showed a flashing green light – clear to taxi. There was no radio talk. No point in letting prying ears know what is going on. Slowly, the aircraft started to move.

'Just to recap, you can apply differential throttles to control direction while taxying, but, as I have said, I recommend not. Keep on using full rudder first then a touch of brake if you must; that way you check that both the rudder and the brakes are working.'

The concentration became intense. The aircraft bumped and trundled over the grass surface, then stopped at a holding point. The pre-take-off vital actions were carried out and the pilot looked to the tower and saw a steady green light – clear to take-off.

'Remember what we said about torque? The engines are both trying to turn you left. You have a critical gap between revving up to start the take off run and the rudder becoming fully effective. To counteract this you need to apply full right rudder and use differential throttle. Twist your fist to lead with the port engine to keep straight until the rudder comes into play. When you have to start taking off full rudder to keep straight you can ease the throttles up to full power on both engines – not before. Lift the tail wheel, then positively back on the joystick so she un-sticks so to speak. It's not so different from the Beau but with the Mossie everything has to be more precise.'

The student complied. It all went 'text-book'. The thumping and bumping stopped as the aircraft became airborne. The undercarriage retracted, the flaps came up and the aircraft settled with the superchargers set to a climb setting.

At the end of a week of fine weather, the training programme completed, the student was sent into a 'solo' exercise. The student didn't actually go 'solo', but the instructor conspicuously unplugged himself from the intercom. and sat with his arms folded across his chest looking intently out of the side window. This was his way of saying, 'Now get airborne again and you're on your own. Don't ask me for any guidance because I won't reply. Try to get us on the ground again with no dramas.'

The flying test was duly carried out. The Instructor was pleased and allowed himself the luxury of a smile. He was particularly pleased about his student's single-engine approach work, which went very well. He had hammered home that mishandling a single-engine approach to landing in a Mosquito was the one thing that could get you into serious trouble.

Apart from the Luftwaffe of course.

Chapter 9

Flogging Willing Horses

While the "boffins" were carrying out their silent arms race interfering with each other's strategic bomber guidance systems, a line of new tactical thinking was emerging. Tactical it may have been, but it moved from the defensive to go on to the offensive.

* * *

The Squadron Leader had been a bomber pilot before being posted for a type conversion onto Mosquitos and afterwards take up a post as the deputy commander of a Mosquito night-fighter squadron based somewhere in the south of England.

His new aeroplane delighted him. The 'Mossie' was a killer. During his first month of duty after conversion to the new fighter he had been involved in four unsuccessful engagements with the enemy. Unfortunately, the last one had taken place in a half-moon, which had caused a glint that caught the eye of the rear gunner of a Junkers 88. The rear gunner was himself new to this dangerous game and was nervous, which meant he chose to concentrate on scanning the moonlit sky rather than drinking ersatz coffee with his mind in neutral. He blinked. To his abject horror he found himself staring straight at an RAF night fighter outlined in the moonlight and in line-astern formation on his own aircraft. Heart in his throat, he yelled for evasive action and opened fire. The RAF fighter-pilot beat him to the draw by only a second, but it was time enough for his cannons to set fire to the Junkers' starboard engine. It was not enough to put the German rear gunner off his aim however, and his machine-gun bullets struck the Mosquito in the front and the belly area. The engagement was broken off. Two of the Mosquito's cannon had been disrupted by return fire and the Squadron Leader had gasped as he felt a sharp searing stab in his left foot. He landed at back at his base assisted by radar steers 30 minutes later. He felt distinctly light-headed and the acuity of his night vision was impaired. In the deafening silence that always followed shutting down the aircraft, he felt a squelching in his left flying boot. Then he lost consciousness.

He regained it again on the following day in the local hospital, where he learned that he had lost some of his left foot. There followed weeks of further surgery and rehabilitation, and the fitting of special left shoes and boots – but the joyful sight was the left flying boot because it confirmed that he would be returning to flying as the Doctor had promised! After all – if Douglas Bader could do it with no legs at all, how important was the loss of half a foot?

During the rehabilitation he was returned to his Squadron and put on ground duties. As this process was ill-defined, he lost no time defining it. He was still in line to be the Deputy Squadron Commander and in initial officer training he had been confidentially advised that if you want a promotion you should behave as if you have already got it. People need to be able to see you as credible in the role. He concerned himself tirelessly with battle reports, technical development and talking endlessly about tactics to his returning aircrews and about the co-ordination of tactics with Ops.

Group Captain Johnny Johnson, one of the foremost day fighters in the early days of the RAF's war, said of night fighting, 'it was a story of the hazards of seeking and stalking in the darkened skies. So entirely different from the quick thrust and parry of our own daylight duals against the Lufftwaffe'.

The Squadron Leader had, many months ago, become deeply aware that what was being attempted was risky and unprecedented. Fighting like this was indeed different. Night fighters hunt like lone wolves in the blackness of night – except that their pilots didn't have night vision acuity of one tenth of that of a wolf. The pilots were pushing the boundaries of what was possible to hunt, stalk and ambush in conditions of near inky blackness and roaring sound. Day-fighter pilots lived under the sun in a unique high speed three dimensional world. If they became disorientated from all this pulling around, bunting, diving, rolling and getting into a spin it was sometimes referred to as 'not knowing your arse from your elbow'. They talked with their hands. The days fighting with the enemy was shared with comrades, over thin war-time beer in the Mess, with wild gesticulations of both hands, one representing you and the other representing the enemy.

'Jerry was on my tail so I pulled up hard (his right hand shoots up to the vertical) and he whizzed underneath me (his left hand shoots down in a dive) – couldn't follow me you see – but there I was, hanging on the prop, airspeed dropping like a whore's drawers on Boat Race week – right bloody Horlicks...'

Night-fighter pilots used to pose a riddle.

'How do you stop a Spitfire pilot talking?'

Answer? 'Tie his hands behind his back.'

Within a short time the Squadron Leader really knew his stuff and the word was starting to get around that this officer was another Douglas Bader type – earnest, over-persistent, convinced he was right, but charming and a damn good thing he was on our side. He was also highly clued up with the latest thinking on tactics. He listened carefully to what the returning crews told him and was keen to listen to their opinions about how to improve them.

The Wing Commander would telephone him with questions such as 'Did you get a conversation with GCI (Ground Controlled Interception) about what happened over Dover to 'Micky' Fynn last night?'

The Group Captain would telephone with questions such as 'Sorry I wasn't with you yesterday – kept too busy showing Royalty around the shop. Now, did your meeting with GCI go alright?. I mean are these new developments going to do the trick. Are we improving?'

'Well the equipment is getting better Sir; and our tactics are becoming more crisp, but we need more progress to exploit the full potential of the system. We're not there yet.'

'Well I have to go to Bentley Priory on Thursday to give an update, so just drop me a memo will you, including further improvements we need to make. Try to keep it as positive as you can.'

But the Group Captain was ordered to bed with influenza on the Wednesday, which is why, in the late spring of 1941, the earnest Squadron Leader found himself 'invited' to report for briefings with a senior staff officer at HQ Fighter Command, RAF Bentley Priory.

* * *

No-one ever managed to discover what had happened to the Junkers 88. The engagement had taken place over the sea and the chances of reliable confirmatory wreckage was as near to nil as made no difference. The navigator had confirmed that the engine did catch fire, but no final impact with the surface had been observed, and the victory could therefore not be confirmed. On closing the battle report, the Intelligence Officer had listed it as 'A Probable'.

* * *

The listener, for the moment, was an Air Commodore, one of the senior staff officers of HQ Fighter Command.

'Well, Squadron Leader, this memo of yours has been forwarded to me by your Squadron Commander. I have asked you to come in because it is directly relevant to ongoing conversations here at HQ as a matter of fact. I would like you to be perfectly candid. I want to see how the view on the front line fits with the concerns we have here at Command. You go first.'

'Thank-you Sir. Well, as you know, only six months or so ago it was still a bit of a shambles trying to get to grips with shooting down bombers at night, but now, things have come together better than we feared it might. The GCI set up is on the ball – they have developed their skills well. Their ability to turn us in right onto a target now has raised our confidence. I can't describe what a relief that confidence is Sir. The equipment in our night fighters has improved and is effective, and we now have some very definite views about tactics. We know what works. This new "cab-rank" development, for instance, where fighter aircraft stooge around a radio beacon waiting to be called in is a big improvement.'

'Bring me up to date on that would you Old Chap – how is that working?'

'When we report on stint we will be told to sit in the crew-room or alternatively to man the aircraft, carry out pre-departure checks, and wait in the cockpit. The order to start-up and taxi comes and those of us at cockpit readiness get going. Crews in the crew-room are then ordered out to prepare their aircraft to replace us on the ground. If we are cleared to take-off we make for the designated holding beacon. If GCI already have trade for us…'

'Have what?'

'Oh, "trade" Sir, that's GCI chat for an incoming bomber stream they've picked up on their radar scopes.'

'I see. Go on.'

'If GCI have trade for us they direct us onto it and set us up in a position where we can take over on the aircraft equipment to attack. If not, we are asked to hold around the beacon and wait until they have. The main tactical objective is to always have two or three fighters over the beacon waiting to be dispatched into an attack. Whatever is going on in any battle, that reserve of holding fighters must never be allowed to diminish. GCI must be excellent at judging when to scramble the waiting fighters up from the ground to ensure that the level of airborne readiness never depletes. The bombers still come across in large numbers, but the tactics have changed for them to operate singly rather than in night loose streams – we just have to slip in behind them unnoticed'.

'Why am a waiting for you to say "But.."?'

'We're not taking the initiative Sir. We wait for the Luftwaffe's next change of their tactics and then we have to start to work out how we might counter it. Meanwhile the bombing continues. So far, eventually, we have been successful but the Luftwaffe is good at this. One day they're going to give us a nut we can't crack early enough and we'll be back in serious trouble again. For instance, their crews seemed to us then to be surprisingly relaxed on look-out before we got the interception procedures right. We knocked them about for a while, but now it's a different matter. They don't just pay lip service to defensive weaving any more – they do it in earnest. They have re-learned that as tiresome as evasive manoeuvres are when you decide to do it as a precaution even when you are not actually being attacked, it's better than having to do it when you are actually attacked because you didn't do it in the first place. Sorry Sir – does that make sense?'

'Perfectly. Go on.'

'Both sides know that the development of electronic equipment is going to be critical in this contest, and both sides are pushing ahead with that as fast as they can. That must continue, but not to the extent that we put all the eggs in that one basket. Tactics are equally important, and frankly Sir – their thinking is ahead of us.'

'Are you saying we are going to lose this phase?'

'Only if we try to deal with this defensively. We have no choice now but to take it to them. We have to stop thinking that the battle only starts when they have crossed the Channel. We have to hit them at their bases as they take off; or before they take off, or while they're still in their pyjamas if necessary. They have already started doing it to us – they have already thought one move ahead of us and now they will be guessing how we might counter it and how they then might make themselves ready to frustrate that counter-move. German intelligence knows where we are. Our night-fighter bases I mean. They know where the squadrons are based that have started hitting their bombers so hard; and they are coming after us, just like they came after Biggin Hill and Tangmere during the Battle of Britain. They can't put our eyes out because they can't hit tiny radar units in the dark. But they know where our bases are and they are going to try destroy us on ground and give their bombers a clear run at industrial targets. The lucky irony in this is that while their tactics are right, they are not fully effective because they have no confidence in them at the moment. It's experimental, so they are trying on a small scale. But they can still find our bigger cities and so continue their strategic objective of industrial bombing. If they desist in the near future that will be their big mistake because the tactics are correct but only if they really go all out to punish us. There is one Luftwaffe crew who regularly try to hit the hangars at Hinchley. It's the same crew – or at least the same aircraft – we have photographs to prove it. Also, there is something wrong with his bomb-sight so he regularly bombs around the same stretch of open fields, his bombs consistently landing short. Causes hilarity in the Mess when everyone has crawled out from under the tables and brushed the dust out of their Brylcreem – but I'm not amused Sir, and I'll tell you why.'

'Two nights ago a German night fighter jumped a Beaufighter on finals approach at RAF Wittlebury. Crashed in flames, everybody gone. A Beaufighter! How did they do that? Because they attacked it when it was vulnerable – either landing when the crew thought it was all over for the night or taking off when your attention is elsewhere. We have to do the same Sir. We have to range out into occupied Europe, hang around their bases or assembly points and hit them where they live just as they, rather feebly are doing to us. But we have to do it harder and with more determination. We won't stop them with purely defensive tactics. If we are to give offensive aircraft the best chance I would want the Mosquito's superior performance. It was originally intended as an unarmed raider, and while the weight of the guns it now carries will degrade its performance a bit it won't be enough to prejudice the operation's success'.

The Air Commodore chuckled.

'Don't ask for much do you Squadron Leader? Very well. I will have words with the AOC in C. You may have talked yourself into it. I will recommend you are appointed to raise these operations and as far as the development of the first proving raids are concerned you will regard them as experimental special operations. I'll assign you some aircraft to start with. I assume given that you are walking with a stick you won't be flying for a while; but that's probably as well. I think you are best used for face-to-face de-briefing of crews, analysis and tactics development. If you feel the need to appreciate things from the air you may fly as the navigator, but only if strictly necessary and I mean that. You may hand-pick your crews. I will clear all this with your Commanding Officer. It's not a usual set up so don't embarrass him. I want written reports after each enemy raid with the salient points summarised, the conclusions drawn, and what aspects will inform tactics. These must be copied first to your C.O., who will, of course, be invited to comment on any aspect. Alright?'

'Yes Sir. Thank you.'

'Well, If there's nothing else have a safe trip back and thank you for coming over.'

'My pleasure Sir.'

The Squadron Leader stood up, placed his peaked cap on his head, saluted, and deliberately not using his stick, marched painfully out of the office closing the door behind him.

The Air Commodore looked thoughtfully at the telephone and drummed his fingers. Five minutes of thought later he picked up the receiver and called his boss, the Air Officer Commanding.

'I've just had a chat with him Sir. I think its the confirmation we were looking for. He's urging the same line we are getting from from Basil – more offensive use of night fighters. (AVM Basil Embry – Fighter Command).He's earnest to say the least but I can't fault his reasoning. He's not one of those who just states the problem – what we mostly know already. Since he was wounded he has obviously immersed himself in finding solutions and he is opinionated. He can jump to conclusions but he also has a high level of tactical appreciation of the problem and seems determined. I think we might have another Bader on our hands.'

'If you're trying to make me nervous David, you're succeeding. Another Douglas? Dear me!'

'Well – he was right about the machine-gun argument to be fair Sir and he wasn't afraid to stick to his guns – no pun intended.'

'Yes he was. What have you told our man to do?'

'When his battle reports arrive, once he has appreciated them and made his comments, to send them via his CO to me immediately. I will send them with my comments to you.'

'Very well – I won't sit on them – they will be back with you same day.'

'I think they'll have to be Sir – I'm expecting a lot of reports. I don't think this officer is going to waste any time.'

* * *

As the months passed, the words of the earnest Squadron Leader were being seriously and urgently applied. Night-fighter Mosquito squadrons were redeployed on 'Ranger' and 'Intruder' sorties initially behind the west coast of Europe, but then deeper and deeper into Europe. Intruder missions were generally against specified targets, ideally to shoot down Luftwaffe aircraft as they took off, or when they were absorbed in training in their airfield circuit, or at their most vulnerable when descending on an approach to landing. Ranger sorties were ordered to fly over enemy territory looking for unspecified 'targets of opportunity' including ships, planes trains or buildings – a list by no means exhaustive. These operations are described in some detail later. There was a more complex version that took place in the skies above Britain however. Unknown to crews of RAF landing aircraft, a protective night-fighter Mosquito might be quietly above them. Their purpose was to give protection, or what came to be known as 'targets of opportunity'. If landing RAF bombers were the prey for intruding Luftwaffe night fighters, in this scenario, the Luftwaffe intruders themselves, would become the hunted.

* * *

The Hauptmann was very bright, in fact he made his superior officers sometimes feel uneasy.

'We have them plotted Herr Oberst – their bases I mean. We don't know the exact position of all of them because they keep changing, but we have enough to make a difference. These are their night-fighter bases, and here, further back, is where they are basing many of their bomber units for the moment. When they come back from a mission they come back to these bases, which is where we will be to shoot them down in the circuit or on final approach. We can calculate to within thirty minutes when they will arrive. We have Junkers '88 night-fighters in holding areas right now and waiting.' His finger circled the areas.

* * *

The higher Mosquito held off at 5000 feet. Below his aircraft, the pilot of a returning bomber, also a Mosquito, settled uneasily into his approach to landing. The higher aircraft, with its pilot and navigator, were the top cover. They were waiting in case a Luftwaffe 'Intruder' to tried to ambush a landing bomber. To that crew him, a returning bomber was the bait. When he located a Luftwaffe night fighter they would pounce. 'Intruding' had proved a deadly tactic whereby an aircraft returning from a raid with an exhausted crew could be caught napping and shot down. So RAF Fighter Command had secretly formed a special duties formation to take care of the problem.

The navigator of the higher Mosquito placed the aircraft well down moon from RAF Hensham, a bomber base. The pilot tuned a radio into the BBC and whistled along to *Whispering Grass*. The minutes slipped into hours. The moon moved in the sky. The navigator kept making up reports to help the pilot keep concentration. Then the radio crackled,

'Dogger Bank 3, Dogger Bank 3, Tether-end, Tether-end.'

The navigator didn't need to consult the code-book. He was being told that in his sector there was enemy aircraft activity, aircraft which had not responded to the ground station's challenging IFF (Identification Friend or Foe) and were therefore deemed hostile, sometimes referred to as 'Bogeys'. They frustrated the ability of the GCI to track them by remaining at a low height. The RAF fighter turned back towards the moon and five minutes later the navigator thought he had caught a glint of light below and ahead to the right. He bent forward in his seat and said to the pilot sitting alongside,

'Bring her starboard 40 degrees would you?'

'What's up?'

'Maybe nothing but keep coming, keep coming, and roll out now.'

Silence.

The pilot asked,

'Did you pick this up as a radar contact?'

'No – too many ground returns – just good old Mark One Eyeball.'

'Oh Dear...we could be here all night then.'

'You really think so do you? Well. Just look down. Two o'clock low – range about four miles.'

The pilot looked and saw nothing. He looked slightly to one side of the reported position to search using his peripheral vision where human night vision is most acute; and then he did. A glint and an outline in the moonlight that was definitely German. The 'bogey' was turning towards RAF Hensham, the bomber Mosquito base. That looked definitely like loitering with intent. Although he could not see a third aircraft he had to assume there was one around and it was on his side. With deliberation he slowly closed the throttles, still with his Mosquito flying in the opposite direction to the 'bogey'. He was too close in to the opposing track. The Merlins were spitting and back-firing, each backfire producing a flame. How could that Jerry pilot miss seeing them? His night vision now ruined, he imagined a line of attack and flew it using instruments only. He turned 45 degrees left to space himself for a turn back onto the line all the while descending. Then came the bit that the navigator detested. As the altimeter reading approached 2500 feet he stood the Mosquito almost on its starboard wing-tip and pulled 'G' with top rudder applied, side-slipping, and flouting lots of rules. He sliced down towards 1000 feet, the height at which the landing Mosquito would be flying the approach to landing less than a mile ahead of the pursuing Luftwaffe intruder.

The turn had worked and in the moonlight it was possible to regain visual contact with the intruder.

The pilot of the Junkers was trying to sort out his own problems. He too had been too close to his quarry when he spotted it and he also had to go into gyrations to position behind it. He picked up his target but had trouble stabilizing his aircraft, still too high but good visual with the exhausts of the landing bomber ahead of him. He concentrated with all the power of a predator. Unfortunately for him he did not know as he closed in for the kill that he too had a tiger on his tail. His BMW radial engines too did not like being throttled back and put out flame flashes from their exhausts. The attacking Mosquito crew, now number three in this line-astern descending queue behind the German, breathed sighs of relief as the aircraft stabilized in a good firing position and they realised that they were going to be able to carry through the attack. The pilot of the tail-end Mosquito switched the armament master switch to FIRE. At the same time the pilot of the Junkers was doing exactly the same thing.

The air traffic controllers sitting in the darkness of their tower peered up the approach path intently. The only lights showing were the darkened runway glims indicating the landing area. In the tower they were showing a steady green light from their Aldis lamp indicating that the approaching bomber aircraft was cleared to land. The air traffic assistant suddenly tensed and said,

'Oh God! There's more than one aircraft on approach! There's two!'

'Hell's teeth – are you sure?'

'No – but there was what looked like an exhaust flash in completely the wrong place.'

The controller was not prepared assume that the second aircraft was friendly. He snatched up the Aldis and changed the signal to a flashing red light – 'Danger – Do Not Land!'

The pilot in the leading Mosquito saw the flashing red light and with the mildest of curses after a long night pushed both throttles to full power and selected the undercarriage and flaps up. The engines roared and the Mosquito surged upwards and away from the enemy gunsight taking the Luftwaffe crew completely by surprise. That split second hesitation cost them dear. The pilot in the following Mosquito opened fire at the Junkers. The Mosquito vibrated violently as its four cannon discharged a three-second burst. The pilot could not accurately judge his range to target, but he found out he was too close as pieces of the Luftwaffe aircraft blew away, one damaging the right side windscreen. The pilot of the Junkers was mortally wounded by the burst and lost consciousness slumping over the control column, which put the aircraft into screaming dive until it hit the ground and exploded. His crew, knowing their lives were now measured in seconds, died in a rictus of terror. In the utter chaos that followed the air traffic controllers had to deal with stray cannon rounds that didn't seem to miss the tower by very much, an explosion so close it burst an external door and broke several windows with flying debris, summoning emergency services to attend vainly the crashed aircraft, and generally trying to work out what had happened. They had seen and especially heard, two aircraft overshooting and climbing away. They had to believe and desperately hope that the crashed aircraft was not RAF, but even that was not clear. It was the following morning before they got a confident description of what had happened. Trying to contact the two Mosquitos who were by then climbing away towards other aerodromes had produced no result. The hunter Mosquito had lost sight of the bomber as he pulled away into a climb and pointed at his own base. Nothing was said at first. Then the navigator said

'Quite a night. All that flying around with nothing going for days and then this. Now it's started are they all going to be like that do you think?'

'Yes I think so – it fitted the job description didn't it?'

'I think you did well Phil.'

'Thanks for saying so; but we are going have to be self-critical Jim. There was too much back there that could have gone really wrong.'

'For sure.'

They joined the circuit at their own base on a flashing green light from the air traffic tower and carried out the pre-landing checks.

'Brake pressures – above 200.

Superchargers – set.

Undercarriage – selecting DOWN, running, and indicating locked.

Propellor pitch to FINE.

Fuel – sufficient – fullest group selected.

Speed – below 160.

Flaps – fully down.

Steady green light from the tower – Clear to Land. Come home, all is forgiven.'

With the flaps and the undercarriage selected, the rate of descent, if uncorrected would markedly increase putting the Mosquito in an undershooting flight-path. This was anticipated and the throttles were increased to maintain it and at the same time preventing the engines' spark plugs from fouling – it was a deliberate manoeuvre.

They had already noticed that there was a thin layer of fog covering their aerodrome. The pilot thought, 'For heaven's sake – anything else?'

He lost useful sight of the glims in the fog in the final 50 feet and closed the throttles, holding the nose 5 degrees above on the Artificial Horizon.The plane thumped onto the grass. They came to a stop, and because they were now at the bottom of a fog-bank they shut down both engines right where they were, unable to see to taxi the plane. Dawn was breaking. They climbed out of the aircraft and knelt on the grass, not to pray, but because their backsides were numb. There were larks singing. And blackbirds. In the distance a vixen called. The engines went crack-crack-crack as they cooled. After shouting and blowing survival whistles, they were found eventually by a searching ambulance crew, and driven carefully to the Operations hut.

'Everything alright Sir?'

'Yes thank you Chiefie. Everything was just fine. No snags.'

'Glad to hear it Sir. The boys are brewing a cuppa cocoa for you. Are you sure you're alright Sir? You look a bit shattered. Guns went off alright did they?'

'Yes Chiefie – very nice.'

It was 5 o'clock in the morning. They nodded off while being driven back to the Officers' Mess.

* * *

The 'Mossie' and the 'Beau' continued to operate side by side until the end of the war and beyond. They shared many things in common – most notably innovative design and production of aircraft that were more than up to the many roles asked of them. It was this astonishing multi-role capability that made both aircraft so remarkable and so invaluable. Twin-engined, as well as carrying a fearsome variety of weaponry that embraced machine guns, cannon, bombs, rocket projectiles, and torpedoes, they could also carry a prodigious fuel load that could keep them in the air for a long time, and for what must have seemed to the crew, in the darkness and freezing temperatures, like forever. Moreover, they had a crew of two, who shared the workload into manageable proportions, which in turn produced higher crew performance.

The original trials had demonstrated that, with the best will and skill in the world, ground radar at its then current stage of development was insufficiently accurate to guarantee placing a night fighter in a firing position behind a bomber. It put you in the 'ballpark' as Americans say; but being in the ballpark was not sufficient to hit the ball. So there was continual refinement in the way the attacks were finalized – refinement that would enable the pilot to *see* the target in the dark sufficiently clearly to use stick and rudder pedals to make the fine final adjustments needed to aim the guns at it and open fire. As noted, while humans cannot see with no light, they can see objects in low light conditions, especially if these objects are lit in some way and reasonably close in distance. Even a country under a black-out order does afford some artificial light. Runways could be lit at their edges by dim lighting, which became known as 'glims'. These gave critical cues of perspective that enabled the pilot to keep the aircraft running straight during the take-off acceleration; to line up accurately with the runway during final approach to landing; to judge the height above the runway during the touch-down flare and landing; and to keep straight during the roll out. Work continued on ground and airborne radar until they became fully effective

At the same time frantic development was taking place on the Airborne Interception radar programme. Attempting to contract the normal lengthy time-to-operational processes, RAF squadrons were earmarked to assist in the project trials with a view to going operational as quickly as possible as at the same time they helped to improve the equipment. It proved a painful process for the young RAF crews. They were developing a capability to counter a threat that had, for them, sprung out of nowhere as the number one priority. The fighters, some of which had been developed from operational bombers, were new in that they had no operational history in the night-fighting role and as yet had to be considered unfit for purpose. They had to be developed until they *were* fit for purpose. This was literally by trial and error, and at the beginning it was mostly error. They had to keep going at it. It often seemed like an act of faith, or to some, an act of desperation, but needs must. There was a war on.

Chapter 10

Rude Intrusion: A Tale of Two Squadrons

Day bombing had been carried out by the Mk IV Mosquito bomber variant up to the end of summer 1943. From that time however and for over a full year the task of day attacking the enemy was virtually taken over from all types by the Mosquito Mk VI fighter-bomber. Of the 7,781 Mosquitos built, about half of them were Mk VI fighter-bombers, by far the most numerous and tactically potent variant.

The reason that the Mk VI fighter-bomber arrived at all was once again to do with Geoffrey de Havilland's unshakable confidence in the development potential of his *pièce de resistance*. As already observed in the development of the first three prototypes, he and his team were opportunist and not risk averse about development in a direction that had not been requested. This was seen as necessary because bombers without defensive guns mounted in turrets was contrary to RAF and government doctrine. De Havilland was willing to put gun turrets on prototypes to keep the Mosquito in the game if the customers' doctrines insisted; but nor was he hesitant about taking them off again because he knew their weight and drag would ultimately compromise his design's success. He was also reported to have been, in retrospect, shall we say highly optimistic about the number of Mosquitos his company could deliver by when.

So, along with all other variants that were produced, the fighter-bomber variant emerged without defensive guns, as always relying on speed and height, in the case of height often the less the better, to enable the aircraft he produced to do what he claimed it would. The formula seemed to be to develop a new variant, quietly, as had always been the story of Mosquito development to date, to minimise political interference and then reveal it to the politicians and to the RAF *fait accompli.* Sir Geoffrey had done this with the development of the reconnaissance and the fighter versions. Much harder for sceptics to hold on to their objections when the latest version was thrilling industrial, military and political audiences with its virtuosity in the hands of his test pilot son, Geoffrey Junior, above Hatfield aerodrome. So, he might have argued, let's not fret about whether the development of this variant will waste valuable time and money that should be spent on your pet projects – my project is here and now, and you can start to take advantage its unique prowess more or less instantly. Royal Flush.

The point made, production lines went into overdrive and turned out Mosquitos at a rate that surprised everyone waiting in line to receive them. The earnest Squadron Leader had got his way, and though he had not been aware of it at the time, he had actually been preaching to the choir. The Air Commodore had been asked to 'take soundings' from some aircrew highly thought of on the front line and he, more than any other, had resoundingly confirmed what had already been suspected – that as soon as numbers would permit it would be a major step forward to go onto the offensive by extending the night-fighters' radius of operations out into western continental Europe. The Air Officer Commanding Fighter Command issued instructions to this effect.

There was dual aspect to these appreciations and the second involved the evolution of what became called 'Intruder' operations. AI radar-equipped night fighters were to be released to hunt over Western Europe by night. Other night fighters would be released for conversion for fighter-ground attack missions. They would have their AI radar taken out and would be fitted with a new form of wireless navigation system instead, to increase the accuracy of positioning and reduce the workload on the navigator. The first of these aircraft was delivered, and the early job of these crews was to use it and develop tactics for its use. They practised operating at low level to avoid detection in the run in to the attack, terminating with a short pull up to aid positive identification of the target and to enable the attack to be delivered in a shallow dive to maximise speed and impact over it. They were mainly industrial complexes or their supporting infrastructure in enemy occupied Europe.

Shortly afterwards they extended their range of attacks to long range strikes to within Germany itself. This penetration of enemy airspace over Europe was highly risky with the more important targets identified by the enemy who then defended them with anti-aircraft guns and fighters. The Focke-Wolf 190, some argue the best single-engined fighter of the Second World War, was a particular hazard. When the plan eventually came together RAF night fighters ranged over Western Europe to intercept enemy aircraft in flight and shoot them down. The fighter-bombers attacked the enemy where they lived – their airfields. These measures were tactical, but offensive. With the passage of time the onslaught built up to continue by day and by night. No rest for the wicked.

The first enemy aircraft to be shot down by an Intruder Mosquito was a Dornier 217 over France. This was a night engagement.

This particular Mosquito was not a night fighter positioning with airborne radar, but located the target by eye, stalked it and duly shot it down. The Dornier pilot had obligingly left his navigation lights on and that was the last mistake he ever made. Days later another patrol over France took the form of moving from one enemy airfield to another and then another and so on. Eventually it paid off when the crew spotted another Dornier which seemed to be carrying out circuit training, and which was shot down. Nothing succeeds like success and the crew decided to return to base via the airspace around Evreux near Paris, where they spotted a Heinkel 111. They shot that down too for good measure.

There were some losses but mercifully few. One sortie had to be terminated when an engine failed and although the pilot made it back to base he lost directional control on landing and careered straight into a steam-roller which wrecked the plane. There are some machines that even a fighter-bomber shouldn't mess with.

The autumn and winter of 1942 brought poor weather, which discouraged the Luftwaffe from mounting both operational and training sorties, and made things difficult and fruitless for the RAF. It also brought home that the contribution of fighter-bombers was not going to be maximised by hanging around France and the Low Countries all the time. There was a demand for the ability to carry more fuel to increase significantly the radius of action of Intruder operations. This was an extension of the argument that the earnest Squadron Leader had made in the first place – pushing for widening night-fighter operations to disrupt what the Luftwaffe had heretofore regarded as safe bases and training areas in France. The argument was that a long-range force of intruding Mosquitos would put paid to any comfort the enemy might be enjoying about being able to maintain safe operational havens if they just dropped back a bit to the east. Continually dropping back to the east might have been a good tactical move but it had important strategic implications to the advantage of the Allies – the enemy was giving away airspace over which it could claim air superiority. This was accepted at Fighter Command, who started the ball rolling by adding extra fuel tanking to thirty of their night fighters to enable them to fly deeper penetrations. Their night-fighting capability was lost to these aircraft because the modifications included replacing AI radar with the latest navigation equipment, including a system called *Gee.* It was also recognised that Intruding was an expanding special operation so training in the skills it required could not remain piecemeal. The various contributing sources were consolidated into a single operational Intruder training unit based in Shropshire. These developments had the feel of permanence – things were changing.

The *Gee* navigation system, when tuned to the radio appropriate frequencies, produced a series of regular low notes in the navigator's head-set. At any point there would be at least three ground stations continuously transmitting. These groups of notes or pulses varied proportionally with range from the transmitter and counting them would therefore accurately indicate the distance of the aircraft from that transmitter. The navigator had to work fast to make this work at all. He carried a chart, which had distances from the transmitters shown as a series of concentric circles with the transmitters at their respective centres. At first glance the map looked like a cat's cradle. The navigator would tune into the first ground station. By counting the pulses within their groups he would know that the aircraft was somewhere on a particular distance arc. He would then tune to another ground station and do the same thing all over again.

He now had distance arcs from two transmitters that intersected at one point making a fix of position. Repeated with the third ground station, he now had three arc lines that probably would not exactly intersect because of system error and due, of course, to the fact that a Mosquito in flight changes position quickly. The three distance arcs would therefore generally comprise a small triangle referred to in 'Nav-speak' as a 'cocked-hat'. The navigator would then place a little dot in the middle of the cocked-hat with the point of his pencil, *et voilà!* Their position had been determined more or less exactly. It was more than accurate – it was elegant! The navigator would gaze at it in admiration. (He then had to break his reverie and remind himself that it was also important to tell the pilot where they were).

Initially, Intruder aircraft were converted night fighters. The MkVI custom-built fighter-bomber was to come later. Intruding was organized at squadron level, with squadrons allocating perhaps three or four of their strength to these sorties. Then came the red-letter day at the beginning of 1943 when No 605 Sqn was earmarked to be the first squadron to be fully committed to Intruder missions. Its modified aircraft started to arrive in February 1943. A second specialist Intruder squadron was earmarked shortly afterwards – No 418 Sqn Royal Canadian Air Force – who commenced operations in May. These two all-but-forgotten squadrons formed a rivalry in terms of scores of enemy shot down. Both furnished Intruding tally-boards and the scores started at once.

Fighter Command had been wise enough to realize that the way to get the best results from these newly appointed free-booters was to tell them what it wanted them to achieve, but not tell them how to do it. Below are just a very few examples of how effective this type of raider could be once honed to a sharp edge and left alone to do what it could get away with.

In May 1943 a Mosquito of No 418 Sqn shot down his third Fw190 of the night, visually by moonlight – he was not AI radar equipped. This third kill was also his squadron's one hundredth victory. Beer was drunk.

After an initial period attacking a range of targets, Intruders were ordered to concentrate on the disruption of airfields and aircraft, leaving transport infrastructure to Bomber Command. In January 1944 two pairs of aircraft of No 418 Sqn took off in the afternoon from RAF Ford, a south coast airfield, the first pair led by the squadron CO. The pairs took off and hunted separately. It was the depths of winter and they didn't have a lot of daylight left, but the crews had hangovers and they went out with bad tempers . On their return a debriefing took place. The air-fights had been furious but '418' had shot down seven aircraft. Even more, an examination of the timing of the engagements showed that by coincidence the shoot-downs had all occurred within a period of 37 minutes.

In February orders were changed again. Both '418' and '605' were to spend a hard month concentrating on shooting down Luftwaffe bombers returning from the expected re-commencement of German bombing raids over Britain, known as the *Steinbock* raids. The Intruders either infiltrated the streams or chose to hang around Luftwaffe bases all over Northern Europe, invisible, inaudible, waiting for them to return. The psychological pressure on the Luftwaffe crews must have been intense knowing that they were liable to be attacked by a Mosquito Intruder any time after they had started engines, by a night-fighter Mosquito during the transit to target, by a night-fighter Mosquito on the recovery back to base, and by a night Intruder Mosquito even on the ground taxiing back to dispersal after a raid. It is thought that this unrelenting pressure at every stage of the Luftwaffe's *Steinbock* operations gave rise to the expression, attributed to Reichsmarschall Herman Goering, '*Moskitopanik*.'

In the same month a pair of day Intruder Mosquitos crossed the Channel in broad daylight. They started things off by strafing three aircraft on the ground at an airfield. Moving to the south they sighted a He 111z glider tug, known as a Bi-Heinkel, towing two Gotha 242 gliders. The two Mosquitos throttled back, and lowering their undercarriages, slowed to the lowest speed they could, then going around, took turns to open fire. The first glider went down, then the second and finally the Bi-Heinkel. They strafed another Gotha on the way home. The Bi-Heinkel was a strange beast having been constructed, rather like the twin Mustang, by blending two aircraft together to make one. The pilots came home and cheekily asked for it to be credited as two destroyed and the pilots would share one each. The request was granted by Command HQ, no doubt to much amusement!

In March came more spikes in a generally mounting tally by both these remarkable squadrons. The examples below are taken from the exploits of No 418 Sqn, but it should be remembered that this squadron was in a friendly but nonetheless determined rivalry with No 605 Sqn, which was racking up a similarly remarkable performance.

In March, No 418 downed twenty-four aircraft for the loss of one Mosquito. A pair of their aircraft, intruding on the Spanish boarder found themselves in a fight with a Focke-Wulf 109 D 'Long Nose', the latest, fastest version of this formidable fighter. They managed to shoot it down, but in the process pieces of the disintegrating aircraft damaged one of the Mosquitos and one engine had to shut down. They flew home with one of the pair on one engine – a track of over 500 miles mostly over enemy occupied territory.

Later in the month two '418' Mosquitos were dispatched on a day Ranger and duly went off towards Switzerland looking for trouble. They shot down seven aircraft early in the sortie and, still fat on fuel, eventually found an airfield with twelve aircraft conveniently parked in a neat straight line.In two passes they strafed them all. All four aircrew of that formation pair were awarded the Distinguished Flying Cross. (DFC).

One pilot of '418' reached the end of his tour and was stood down and told to go on leave for a rest. Battle fatigue was becoming recognised as a medical condition that had to given serious consideration. Not for everyone apparently. Instead of heading for London, this pilot could be found still hanging around the squadron huts asking for any flying going. Eventually, his dogged persistence was rewarded and Ops gave him a day Ranger to fly. He headed towards the Loire valley and, as the sortie was a Ranger, did so for presumably for no particular reason. He spotted a Fieseler Storch a low speed army co-operation aircraft with a very low stalling speed. It could be thought of as the German equivalent of the British Westland Lysander. The crew of the Storch realised they about to be attacked, promptly landed in a field, and ran for it. The Mosquito pilot destroyed the Storch and pulled away towards Lyon, where he destroyed two further aircraft before turning back towards base. Shortly after settling into the cruise he spotted two Henkel 1-11s, throttled up again, engaged both and destroyed those too. It was a strange way to spend your leave, even for aircrew, but the pilot, who

Armourers prepare to load four 500-lb MC bombs into the bomb-bay of De Havilland Mosquito FB Mark VI of No. 464 Squadron at RAF Hunsdon, Hertfordshire. The rack of 4 Browning 303mm machine guns can be seen protruding from the nose. The muzzles of the 4 Hispano cannon also be seen in ports in the fuselage below them.

Pin Up Boy. The Mossie caught the public imagination with its speed, altitude performance, and sheer good looks. They called it "The Wooded Wonder".

Group Captain P C "Pick" Pickard (centre), Commander of No. 140 Wing, No. 2 Group, intrepid leader of Mosquito special operations, at Hunsdon, Hertfordshire, prior to a daylight raid against flying-bomb sites in the Pas-de-Calais. He was killed on the Amiens Raid.

Air-to-air photograph of a deHavilland DH.98 Mosquito B Mk. IV Series 2 bomber of 105 Sqn. of the Royal Air Force in flight. 1942–3.

De Havilland Mosquito B.XVIs, including the Percival-built PF563, closest to the camera.

De Havilland Mosquito B Mark I of No. 105 Squadron RAF.

Mosquito FB Mark XVIII of No. 248 Squadron RAF Special Detachment Portreath, Cornwall. The 57mm Molins gun mounted underneath the nose is clearly visible. The Mollins Gun was designed as an anti-tank cannon firing one 6lb shell a second. Mounted in the Mosquito they were a fearsome weapon system. The version is sometimes referred to as " Tsetse".

De Havilland Mosquito PhotoReconnaissance Mk XVI of No. 544 Squadron RAF based at Benson, Oxfordshire, December 1944.

Wing Commander H I Edwards VC (left) and his Navigator approach their aircraft before taking off on Operation Oyster, a large multi-type formation ordered to attack the Philips Factory in Eindhoven, Holland.

"The Gestapo Hunters", a painting by Michael Steele Morgan, signed by aircrew who flew the raid. It depicts a 'vic' or three Mk2 Mosquitos exiting from the Aarhus raid. (*Image by kind permission of the SFC, London*)

was an RCAF Flying Officer called Charlie Scherf made a career of it. When he was finally, firmly and definitely stood down from '418' it was with a score of twenty-three enemy aircraft destroyed, of which, it transpired, all but six were achieved during periods of leave spent hanging around Squadron Ops cadging missions to fly and generally making a nuisance of himself.

Most scoring was done during daylight Intruder sorties. No 605 Sqn remained hot on the heels of '418', but found itself restricted to night Intruding, and so never caught up with its rival. Their record was every bit as admirable. Shortly afterwards their CO, Wing Commander Hoare reached the end of his tour and was stood down with his score standing at thirty-three aircraft destroyed and thirty-five damaged on the ground by strafing.

In May 1943, the new custom-made Mk VI fighter-bombers started to arrive, and No 418 Sqn was the first to receive them. This was not a modified night fighter, but a bespoke marque in its own right, and as noted, more machines of this marque were built than of any other. This is testimony to the importance of this variant as by that time in the war, in contrast with what went on in the earlier years with Sir Geoffrey doing as he thought best, Mosquito marques were now being constructed to customer demand. The designers had carefully to consider the stated role requirements and build the capability in from the start of construction. There was sensible consultation with some of the front line crews who would probably have to fly them. The 'earnest Squadron Leader' was to have more bites at the cherry. They took the view that bombs were not necessary, envisaging a weapon system built literally and metaphorically around the rack of four Suiza 20mm cannon and four .303 heavy machine guns. Sir Geoffrey, considerate to a fault, gave them the bombs anyway. The Mk VI would be able to carry four 500lb bombs without having to trade away any of its eight guns. It could carry two external wing drop tanks made out of reinforced paper, which extended its radius of action (out and back) to most of Europe. None of these, nor most other Mosquito performance increases could have been possible were it not for the continuing improvements in power output of the Merlin engine by Rolls Royce engineers. Stages of supercharging, water injection and cylinder pressure were just some of the ways employed to boost power output that enabled the capabilities of the Mosquito airframe design to be constantly pushed to greater things.

By May 1943 Intruders were still flown but with lessening intensity with the general redeployment and repurposing of everything in preparation for the D-Day landings. Mosquitos would still carry out special functions such as the interception of V1 flying bombs, which they did brilliantly, but priority was refocused on the prosecution of front-line advancement to the east and making tatters of the German retreat. After D-Day it became increasingly useful to use liberated European airfields either as refuelling points, or potential diversions, or moving over the Channel to use them as bases themselves. Airfields that once had been targets for Intruders now became home. Eventually all of Europe was brought within the radius of action of Mosquitos flying out of various bases. As a '418' last hurrah, two Intruders flew from Belgium to north Germany shooting down no less than ten warplanes in that one sortie. Not to be outdone, a singleton Intruder from '805' destroyed five Dornier 25s in a single sortie.

How could this be? Possibly because confident, skilled veterans, flying high quality machines, were in high-speed combat against 'newbies', who hardly knew how to cope with complicated, powerful fighter aircraft, let alone win.

Reich-dominated territory was shrinking and so were target opportunities. '805' and '418', their free-booting days over, were redeployed to join the fighter-bombers in the newly formed Second Tactical Air Force (2TAF).

Chapter 11

Bouncing Bombs

The Royal Air Force was formed on April Fool's Day in 1918, immediately after the First World War. It comprised an amalgamation of the Royal Naval Air Service and the Royal Army Flying Corps. In its one hundred plus years of history its most famous squadron became, and remains, No 617 Squadron – 'The Dam Busters'. Its Squadron Commander was Wing Commander G.P. 'Guy' Gibson, DSO, DFC, later VC.

The squadron was formed originally as a special duties unit of hand-picked crew and trained initially for a single purpose – to carry backwards rotating cylindrical mines which, when launched at their targets – Ruhr dams in Germany – would skim like a pebble across the surface of the water to strike and destroy them. Some pebble. This top-secret operation was known by the codename *Operation Chastise*, and the aircraft they were to fly was the Avro Lancaster four-engined heavy bomber, the best four-engined bomber on the RAF's inventory and arguably of the Second World War.

Much less well known is that another similar special duties squadron had been formed at the same time. Exactly twenty-five years to the day after the inauguration of the RAF, No 618 Squadron was formed on 1 April 1943 under the command of Squadron Leader C.F. Rose DFC, DFM. This also comprised hand-picked crew trained for a single similar purpose – to carry backwards rotating but smaller spherical mines, which, when launched at their target – the German Battleship *Tirpitz* – would also skim across the surface of the water to strike its flank, roll themselves down the hull to a calculated depth and be triggered by a hydrostatic pistol. The aircraft they were to fly was the de Havilland Mosquito, the best twin-engined bomber on the RAF's inventory and also arguably of the Second World War.

Both weapons had been developed by Dr Barnes Wallace concurrently, the scientist who invented what became known as 'the bouncing bomb'. The anti-dam weapon was code-named *Upkeep.* The weapon to sink the *Tirpitz* was code-named *Highball.* The novel principle both versions had in common was the discovery that the destructive power of an explosion could be multiplied several times if the detonation took place against the side of the object to be destroyed while underwater. The stone walls of the Ruhr dams and the steel sides of the *Tirpitz* qualified in both respects. The backward spinning mines would not only bounce across the water, but on hitting the side of the target would propel themselves down against it to a pre-determined depth whereon it would be triggered by a pressure detonator. The Lancaster carried one such mine per aircraft. The Mosquito carried two, one behind the other. Both weapons were slung externally beneath the fuselage.

In the case of inland dams, aerial delivery of the sort of heavy weapon needed, was the only option. In the case of an attack against a seagoing vessel, surface attack and sinking by one or more surface vessels might serve, as had been in the case of the *Tirpitz*'s sister ship the *Bismark,* sunk by heavy units of the British fleet. However, the *Kriegsmarine* increasingly followed the strategic doctrine of 'fleet in being', whereby a smaller but powerful fleet that remained defended in harbour behind the Kattegat, or concealed deep in the upper reaches of a Norwegian fjord, was a threat nevertheless. They were able to keep a large number of warships of the Royal Navy out at sea looking for them without actually exposing themselves to the risks of battle. Assuming it was possible to find them, how then would they be possible to attack against the surface booms and submarine nets that would inevitably be deployed to keep the ship invulnerable?

The answer again had to be an aerial attack. Unfortunately, a conventional bombing raid from high overhead would have to get sight of the target first, and the Germans had the moorings of the *Tirpitz* surrounded by a large number of smoke generators that could get the ship out of sight under smoke at any warning of impending attack. Mostly though, low cloud was doing that for them.

The possibility of attack by a small force of high speed aircraft from the side at low level at first must have seemed unlikely. Rocket propelled projectiles might get through but damage inflicted against a heavily armoured capital ship might be minimal and quickly reparable. This was a ship that did its job by *not* going to sea and therefore retained full access to repair facilities. And approaches to the ship would be clustered with anti-aircraft guns, quite apart from the battleship's own formidable secondary armament. Losses would be high for little achieved. The projected ratios were not good.

Barnes Wallace remained convinced about the Mosquito and the *Highball.* Two aircraft armed with four *Highball* mines. To make the profile work they would have to drop down to 50 feet at high speed releasing them at less than a mile, then break off and try to get away. Even if the planes didn't make it out, the bombs would be on their way and from that point unstoppable. The chances of the planes getting out were not good. *Operation Chastise* would be executed at night. Attacking the *Tirpitz* would have to be done in broad daylight.

The trials were exciting. No matter how careful you were inclined to be, delivering a heavy weapon at 50 feet and 250 miles per hour was a hazardous business. Progress towards perfecting the weapon for operational use was slow. The test weapons had the habit of breaking up in flight or when they bounced. The programme to increase their robustness during the delivery of the attack was done bit by bit, by trial and error, adding to the thickness of the surrounding steel casing at each stage and changing the backward rotational speeds until the optimum was found. The reliability of the release gear caused problems and had to be improved. As with 617 Squadron, the flight testing and crew training comprised a prodigious programme of flying.

Gradually, it became evident that there was another consideration lurking that was likely to throw doubt on the operationalisation of the project. As is not uncommon, each time a new phase of any trial is unsuccessful, the detractors have another opportunity to detract. The new consideration was a powerful 'back-handed compliment'. The trials were allowed to continue despite set-backs depending on the continuing view that the secrets of this brilliant, literally off-the-wall, new weapon would not fall into the hands of the Germans. It was unlikely that they would work out its nature following the *Operation Chastise* night attack on the Ruhr dams.

However, a broad daylight attack was a different matter – it would be observed, and the secret would be out and up for duplication. The future prospect of the Luftwaffe in possession of a similar technology could not be countenanced. The Royal Navy presented far more capital ships than the *Kriegsmarine*, each one potentially vulnerable to this novel form of attack. If this is so, this would have been a clear example of the need at all times to focus on strategic considerations as, with few exceptions, to be of more importance than the tactical.

An alternative raid was planned using Lancasters. After much searching, the ship was spotted by a photo-reconnaissance Mosquito hidden within Tromso fjord. The ship was attacked in *Operation Catechism* in November 1944 by twenty-nine Lancasters armed with Barnes-Wallace designed 12,000lb Tallboy earthquake bombs. The bombs scored direct hits and near misses, which were probably just as damaging. Over a thousand *Kriegsmarine* personnel were killed and the ship capsized. The threat posed by the *Tirpitz* to Royal Navy heavy surface vessels and the merchantmen that were keeping Britain alive, was at an end.

No 618 Sqn underwent training for maritime operations as a preparation for shipping out to join the Pacific War. They reached Australia and waited for a long period for orders to mobilise. Instead, they eventually received an order from Admiral Portal of the Pacific Fleet to disband.

Chapter 12

Targets of Opportunity

As already described, the contribution made by Intruding operations were initially based on more aggressive use of night fighters, which had their range extended so that some were free to patrol the western skies of Europe out to a radius of action (out and back) of at least 600 miles. The tactical retreat east imposed on the Luftwaffe by the Intruders' presence delivered strategic reward to the Allies – a substantial increase in airspace over which they could claim, with increasing confidence, the air superiority that would be vital for the success of the expected invasion and liberation of Europe. The AI radar was removed from the night fighters and replaced by the *Gee* navigation system so the converted Intruder could no longer be used to carry out radar interceptions at night. There were enough radar equipped night fighters remaining to continue providing effective defence of Britain's night skies.

Serious moves towards assembling a substantial fighter-bomber force for Intruding began. Almost at the same time, another innovation to the Intruder concept was emerging. This entailed the possibility of assigning fighter-bombers to investigate *possible* targets with a view to identifying what they were and once identified, attacking if appropriate; or sending them into areas with no assigned targets at all to see if they themselves could find a 'target of opportunity' worth destroying. Such missions were called 'Rangers'.

* * *

'Any inkling at all?'

The navigator was a Flight Lieutenant just turned twenty-five years old.

'It might be shipping. That's why we've got RPs on. That aside, we can pick and choose.'

The pilot was a Flying Officer, aged twenty-three.

'Funny way to run an Air Force. I'm a navigator – it's not unreasonable to expect to have a target for me to navigate to. It's what I do. All this vague stuff gets on my nerves'.

'Don't let it – I still need you to get us home – that's often the tricky bit – remember?'

These two young men comprised an expert team despite the fact that one of them couldn't even grow a decent moustache yet. They were bonded by the 'Ops' they had already flown. They were what might be referred to as a 'cockpit marriage' – it all went well when they were busy, but when they didn't have enough to do and there was tension in the air, the banter could get a bit pointed.

'Oh they must have said more than that!'

'Just to prep. for a start and watch the air traffic tower. Strict radio silence; we can listen out but no transmissions. All I was told. '

A grumpy navigator unstrapped from his seat and opened the hatch.

'I've got as far down the checklist as I can here. Mind if I shove off twenty yards and have a last leg-stretch – I have a strange feeling this is going to be a marathon.'

They had been brought to cockpit readiness. They were waiting for a radio call or a signal, the Mosquito already prepared. The navigator was not supposed to be twenty yards away doing PT, but a long time at readiness might involve a need to answer the calls of nature, or getting out to ease leg cramp or bum ache, so it was a rule sometimes honoured in the breach. The pilot remained strapped in his seat, radio headset on, listening to the radio traffic. Waiting. By sheer chance, on this occasion they did not have long to wait. Three NCOs suddenly sprinted out of the dispersal hut. The navigator saw them, hauled himself back into the belly of the aircraft and slammed the hatch shut. Already the air traffic control tower was pointing a flashing green light at them – 'Clear to Taxi'. This was not officially a 'scramble' but something was afoot.

Pre-start checks completed, look at the sergeant. The sergeant points to the port engine with his right index and makes a rotating motion with his left hand. The sergeant has a fire extinguishing crew with him. If the engine develops a start-up fire he stands between you and immolation. Sergeants are important people. Fuel cocks on, fuel pumps on, press the start button. The port engine coughs, hesitates, screeches agonisingly, then fires and roars. No signs of fire. Set the throttle at 1,200 rpm. While the pilot is settling the engine down and checking his system gauges, the sergeant walks to the starboard side, points to the engine making a rotating motion with a finger. Fuel cocks on, fuel pumps on, press the tit. The starboard engine goes through the same painful hesitant rituals, then fires and roars. Set the throttle at 1,200 rpm, check the gauges, no signs of fire. Both crew-members are now very busy. Checks, checks, checks.

All ready. Lots of thumbs up.

The ATC tower is now flashing two letters at them in Morse code – 'NE'. This to indicate the initial track after take-off (north-east) while maintaining radio silence for as long as possible. The navigator goes straight back into his drift and groundspeed calculations and checking optimum altitude to maximize range and groundspeed. The pilot doesn't wait. He looks at the sergeant and indicates 'Chocks Away'. He releases the brakes, inches forward the throttles, and slowly, propellers roaring, D for Dog starts to roll. The sergeant stands to attention. You nod, this gesture meaning 'Thank-you Sergeant.' They bump along taxiing distinctly more quickly than normal. Well before reaching the holding point the ATC signal lamp changes to a steady green light – 'Clear for take-off.' With the briefest of pauses to check the brakes, full and free range of the flying controls, both magnetos on line and confirm that both engines were within the normal operating temperatures (they weren't but the pilot wasn't going to quibble), the Mosquito is turned onto the runway heading. It was early afternoon.

'Initial heading north-east then. I'll give you an accurate one when Ground Control lets me in on the secret. Climb to Angels 20.'

'Will do.'

The pilot puts on full right rudder and advanced the throttles slowly leading with the port engine to counteract any tendency to swing to port, which she would if you didn't treat her nicely. At 60mph the rudder starts to bite and he goes to full power on both engines. He pushes the joystick forward to lift the tail wheel and at 90mph starts to ease back on the stick. One last bump on the ground helps the

Mosquito lift herself into the air and within seconds it all comes together. Wheels up. Flaps up. She accelerates like a Blower Bentley, but curiously, the pilot holds her low at 1,000 feet. The navigator, head down, waits for a change in engine note that doesn't come. He glances up at his altimeter, still reading 1,000 feet, and says

'We can't do this low level. We need to get upstairs. Oh I get it! We're going to impress the lovely Marilyn aren't we? OK – let me know when we're ready to rejoin the Second World War.'

The Mosquito built up speed and sounded gratifyingly dragon-like as they passed a satellite relief-landing airfield on their required track north-east of their base. The impossibly pretty Third Officer Marilyn, WRNS, was duly impressed, the stick was pulled purposefully back, speed was converted to a zoom climb and the superchargers set for a normal cruise-climb as they passed 8,000 feet.

* * *

The track across the North Sea had been uneventful. The weather was fine, clear visibility with about half cover of tactically useful cumulus cloud. The navigator's impressive skill with a Morse key and wireless headphones had enabled him to make contact with an operations centre, receive a short coded message, and decode it. It had given four position co-ordinates that when dotted on the chart comprised the four corners of an oblong of a patrol area that turned out to be to the north-east of the North Sea around the approaches to the Skagerrak Straight. He drew the box on his map, took a fix of the Mosquito's position, quickly drew a line representing the required track, applied wind effect to resolve a heading and spoke it to the pilot. Two hours later they were in the box, throttled back to endurance speed to extend time on patrol. After a coded signal indicating they were on station, a further message was received, which the navigator decoded.

'Given we are over the drink, it will not surprise you to know we are looking for a ship. No more details than that'.

'No problem – I know a ship when I see one. Can this particular ship fire back?'

'You know everything I know'.

They flew on, throttled back to endurance speed, scanning the sky and the sea for the next two hours, trying to find the ship that a Royal Navy submarine had spotted while on patrol in the approaches. It had been spotted into sun, in and out of sea mist and through the periscope could not be identified. Sitting just behind and to the right of the pilot, the navigator checked the fuel. Mosquito Ranger crews were adept at extreme range calculations, and taking into account all factors, at least those of which they could be aware, calculated time left to search before starting the return to base. He glanced at the compass.

'No further right please or you'll be pointing us at Sola and they're watching us. If we get twenty miles closer we might not be able to out-distance their Focke-Wulfs if they decide to have a go. You can bet they're just hoping we'll make a mistake like that'.

'Ok – put us forty miles outside their radius of action, just to annoy them. How long before we have to set heading for base?'

'Less than three hours, but don't worry – I've picked up some weather reports. If we have to stay longer the weather over north Scotland is gin clear. How about a beer at RAF Banff?'

'I've heard the beer at Banff is pretty watery since the war started'.

'Lead me to it Boy – I'll suffer.'

The broken cumulus cloud was providing them with cover. The pilot was flying a wide-spaced creeping line ahead search pattern and at every turn he pushed the nose down to drop into the clear to have a good look around at the sea as the aircraft turned. Suddenly, as he passed down-sun he thought he saw a black shape on the glittering surface. Just as suddenly the plane flew back into lower cloud and the shape was lost.

'I just got a contact, or at least I think I did.'

He was cautious – sun on a sea without a clear horizon is full of optical illusion.

'What was it?'

'Don't know – big – so it could be a tanker.'

'Yes, but if it's big it could be something else as well.'

The Mosquito continued its descent through the lower cloud, its two Merlin engines spitting and complaining at being throttled back.

Some time before, a crew based at RAF Eglinton near Londonderry in Northern Ireland, boarded a Sunderland maritime patrol aircraft of RAF Coastal Command. The massive flying boat had taken off from Lough Foyle under orders to patrol the Skagerrak approaches looking for a reported vessel. The Short Sunderland, built in Belfast, could prowl on station around the clock and could carry a heavy war-load of bombs, flares, mines and depth-charges. Coastal Command Head Quarters (HQ) had been informed of the endurance of the on-station Mosquito by Fighter Command and requested a Sunderland be made available to take up the so far fruitless search being carried out once the Mosquito was forced to leave the area to return to base. Having flown over Scotland, they coasted out over the North Sea abeam RAF Turnhouse, the fighter base intended to protect Edinburgh from attack. After coasting out, the radar operator picked up a radar return, reported it to the pilot, who turned to close with it. It was not the same vessel that the Mosquito crew had caught a glimpse of. The Sunderland crew had no way of knowing this – and went to action stations.

* * *

The Radar Operator of the German Cruiser SMS Krieger reported to the Officer of the Watch.

'There is a blip Sir. About 150 kilometres away to the south-west. If it's a Sunderland and he picks us up he could be on us very quickly'.

The Officer studied the blip for a minute, convinced himself that the aircraft was closing, (which it wasn't), picked up the intercom mike and pressed the klaxon. 'Action stations! All guns to south-west! Standby!'

'Here's another smaller blip Sir – to the south-east here at about 40 kilometres. Looks quite fast.'

'Yes I see it. It has to be a Junkers 88. The English wouldn't be stupid enough put a bomber that close to Sola. Focke-Wulfs. Anyway, keep your eye on them – especially that big blip and give me running information of its range.'

The radar operator, incredulous that under such circumstances this officer could live a life so devoid of doubt, so filled with certainty. He presumed that was what qualified him to be an officer. He paused to wipe a pair of tired eyes, then duly focussed on reporting on the big blip. When he eventually remembered about the small blip and searched for it in the scope, it had disappeared.

* * *

The Mosquito pilot increased engine power, pushed the Mosquito into a shallow dive, pulled down his goggles and tightened his seat straps. He selected the Armament Master Switch to RP SALVO (Rocket Projectiles) and scanned forward from side to side. Nothing. Nothing. Then he cleared a thin layer of strato-cumulus, and there it was. A German flak cruiser! Holy God – everyone's worst nightmare! He throttled up to full power and pushed the nose further forward putting the Mosquito into a 40-degree dive, and let her scream. The German sailors heard her, but it was too late. Their guns were pointing in the wrong direction. They were swinging around as the pilot pressed the firing button and the RPs went off in salvo. They destroyed the cruiser's superstructure and the forward guns. Fourteen sailors and two officers were killed. Many more were seriously injured. The captain, who had been conferring below heard a massive series of explosions above his head and raced up the ladders to witness mayhem on his decks. The Mosquito was fast disappearing in a hard right turn, but the cruiser's rearward cannons caught it and, eye-shooting desperately without the sighting system, blew off two feet of his starboard wing tip. Inside the cockpit the crew felt and heard a hammer blow, a violent yaw to the left and saw a ragged end to the starboard wing. After five seconds of fearing the worst, the pilot discovered that she was responding to the controls and he had little trouble righting her. For some reason she still seemed reasonably balanced and after fiddling with the trim-tab controls actually persuaded her to fly straight and level. He tried to control his breathing. Gulping air made you light headed and he needed to think clearly now.

'We still in business?'

'Yes I think so.'

'What's the plan?'

'Well as far as I can tell, we have two good engines and a flyable plane. We are going for Scotland.'

'OK. Steer 240 degrees. RAF Dyce looks closest by a tiny bit, but they have that bloody Haar sea fret at this time of year. Let's see if we can pick up a beacon. Wind is from the south, quite strong, so we'll steer the track and deliberately let her drift north. If there is fog on the Aberdeen coast we have Kinloss and Lossiemouth in the Moray Firth – they're usually clear. I'll confirm that now.'

'Well done Brains – we'll do exactly that. When you've finished anything that, anything else to do?'

'Yeah – I need to finish this Times crossword.'

They both cackled.

There was then a long period of silence. The navigator peered over the pilot's shoulder watching the instruments and particularly the fuel depletion rate to make sure that there was no damage to the fuel tanks or the fuel lines. The aircraft was not difficult to control despite the damage to the starboard wing tip. There was a slight tendency to role to the right but easily controlled with left aileron. This puzzled them both because they would have expected to have required more starboard rudder with two feet of right wing missing. They discussed it, but decided for whatever reason it was one gift horse that should definitely not be looked in the mouth.

Battle shock is a treacherous thing. When you are pumped to the brim with adrenaline and thinking so fast you leave yourself behind you are never more alive. The price you pay afterwards is that your body needs a break and, incredibly, you can find yourself wanting to go to sleep. The pilot bit his lip to resist it, touched his mouth and saw blood on his glove. The cockpit was cold, which helped, but their track was into a late afternoon lowering sun, directly in his eyes, which didn't. From 10,000 feet the sea looked like pewter. From up there it was not possible to detect the restless movements of a moderate sea state – it looked frozen. The pilot, who attended a grammar school near Esher, had specialised in Geography and English literature and had caught a love of poetry. Mesmerised by the brightness of the sky and the burnished surface of the sea, he found himself thinking words from '*The Eagle*' by Alfred Lord Tennyson.

'Close to the sun in lonely lands
Ring'd with the azure world, he stands.
The wrinkled sea beneath him crawls,
And like a thunderbolt, he falls.'

Somehow it seemed appropriate.

Eventually, the Moray Firth started to reveal itself as clear as a bell, so the navigator crossed RAF Dyce out of his deliberations as the sun sank towards the Grampians.

'Hello Kinloss, this is D for Dog – request landing clearance.'

'Hello D for Dog – Kinloss here. We have been expecting you. Clear straight in. Call final'.

Pause.

'Kinloss D for Dog. Final.'

'D for Dog, you are clear to land.'

Steady green light from the ATC tower in case of radio failure. The air traffic controller surveyed the approaching fighter-bomber with his binoculars. And then quietly said,

'Oh No! He's got a hang-up. One of his RPs hasn't fired off". He pressed the button and the sirens screamed. The fire engines and the ambulance crews ran, hearts racing, to man their vehicles.

'D for Dog you are clear to land. When you come to a halt please turn off the runway, point your aircraft north and shut down. You appear to have a hang-up under your starboard wing. Do not taxi back to dispersal. Get away from the aircraft as quickly as you can'.

The crew looked at each other. The navigator said,

'Oh bloody charming!'

That explained why the plane stayed in rough balance with two feet of wing-tip gone. It was the drag of the weapon still attached under the starboard wing. The landing was text-book; they turned off the runway and shut the engines down. As the brakes locked, the hatch came open and in quick succession the two dropped onto the grass, picked themselves up, released their parachutes and, flying boots regardless, achieved two personal bests sprinting away from the aircraft.

* * *

The German cruiser docked at Kiel. The wounded were raced off to hospital. The dead were respectfully taken off the ship. The crew set about their business refuelling, rearming, and making do. Engineers came aboard with their oxy-acetylene torches and steel plating to repair battle damage. The captain went ashore to submit his battle reports via a lieutenant in the Operations Room. By coincidence the two servicemen knew each other well having been to the same school in Wilhelmshaven . They exchanged courtesies, then smoked cigarettes.

'You seem a bit down Sir?'

The Captain said,

'I lost fourteen sailors and two officers Hans. Yes, you could say I'm a bit down.'

The Lieutenant said, 'I'm very sorry Sir, but I'm not surprised'.

'Why aren't you surprised?'

'We're making too many tactical mistakes Sir, which the RAF is not. We need those Junkers 88s back. The ones they pulled out of Denmark.'

'I think the Luftwaffe was told to move them closer to Hamburg.'

'I'm not against protecting Hamburg Sir, but dominating the Skagerrak is of critical strategic importance. We need those planes back in Aalborg, or at least somewhere in north Jutland. '

'We have the Fokke-Wulfs at Sola.'

The Lieutenant could barely hide his exasperation.

'Speak freely.'

'We sit here and watch this regularly Sir. We are up against something new with these Mosquitos. They only tend to come over here when there is cloud cover. They are using that cloud cover to frustrate our interception attempts. Focke-Wulf 190s are single-seat interceptors with no radar. Mosquitos are long-range heavy fighter-bombers with a huge fuel capacity. This gives them all the options they need. Firstly, they have the range to reach us and beyond. If they want to go beyond us deep into the Kattegat they will climb to very high altitudes; when they are about to enter our interception radius they go into a shallow dive at full power and they're through us and gone before we can get to them. That fuel load also gives the option to loiter over the North Sea outside the radius of action of our fighters. They will be acting on intelligence tip-offs, possibly from submarines. They will search for a vessel and when they detect it they will make an attack, and then break off and go home and have another go tomorrow. It's all very efficient – operationally I mean. They can get all the way home for instant maintenance and rearming. They are damned fast. And if they take damage they have two engines and stations closer at hand in Scotland. Won't take them long to set up more Mosquito bases in Scotland either, which will improve their range and endurance capabilities even further."

The Kreigesmarine officer sat down to consider. He raised his head and looked at the display chart.

'What are these marks here on the chart Hans?'

'British submarines Sir.'

'How do you know they're there?'

'We have to assume they are always there on patrol. Strategic logic if you like. If it's what we would do if the positions were reversed, we have to assume that the British are doing it now. Look at this area here for instance. Out here you see – up to two hundred miles or more out, if the positions were reversed that's where we would put a patrol line for maritime Focke-Wulf Condors. We have to assume they are doing the same with Sunderland flying boats or heavy bombers – something with similar patrol endurance – certainly submarines, which can loiter for weeks. Sunderland's can stay airborne for sixteen hours I am told. If they spot something they can attack it. But if they pick something up, say on radar, when they are critical on fuel, they just get on the Morse key back to London and report it. Then they call in the Mosquitos from whereever. With Junkers 88s in Aalborg we can dominate the western approaches here. We need aircraft with similar capabilities to their Mosquitos to get to grips with them.'

Hans paused, put his forehead into the palm of his hand and sniffed. The Operations HQ was now silent and listening to him. So was his superior officer.

'Twenty-five years ago the role of the British Navy was strategically critical; they blockaded us behind the Skagerrak and starved us out. They are going to try to do it again Sir.'

On the following day the captain saluted his admiral and was invited to sit down.

The Admiral said, 'I am truly sorry about what happened.'

The captain said, 'Thank you Admiral. I wish to speak to you about how that might be prevented from happening again.'

* * *

The Mosquito crew had finally caught their breath after their impressive sprinting and were walking the rest of the way towards the Operations hut. The armourers were at a distance from the starboard wing, flat on their stomachs playing a light under it as the sky darkened, trying to get some sort of clue as to whether the hung up RP was in a condition safe enough to approach and disarm it. There was a deafening explosion. The wing of D for Dog went up in the air. The two officers nearly jumped out of their skins, wheeled round and stared back in shock. The ambulance that had escorted their landing quickly moved in to pick up the two armourers who, miraculously, sustained only superficial injuries but had a ringing in their ears for several days. The 'fire-wagon' moved in to deal with the fire that was now burning away the rest of the wooden airframe.

The two crew were warmly welcomed in the Officers' Mess, congratulated and slapped on the back by resident officers including the Station Commander. They drank quite a lot of watery Kinloss beer. The batman was instructed to let them sleep in. Early the following afternoon an Anson – a light transport aircraft – arrived to take them back to their base. They were given the next day to de-brief, furnish reports and take it easy.

On the following day they were back on operations tasked for another Ranger sortie, ordered to get airborne and, if nothing else, find a target of opportunity.

Chapter 13

The Oslo and Amiens Special Operations

The Mosquito marque first to prove itself effective at precision low-level raiding was the Mk IV B bomber version. The general public in Britain were made aware of the existence of the Mosquito by an announcement followed by a feature article and photographs of bomb damage in *The Times* in September 1942. The article described what has gone down the halls of fame as the 'The Oslo Raid' The attacking force comprised four aircraft lead by Sqn Ldr D. Parry DSO, DFC. Their brief was to destroy the Gestapo Head Quarters in Oslo. They took off from Leuchars on the east coast of Fife in Scotland to fly low level across the North Sea to the capital of the now enemy-occupied Norway. They entered Oslo Fjord at 300mph at 100 feet in two battle pairs, but as they approached the city they were bounced by a flight of three Focke-Wulf 190s. The wingman in the second pair had fallen behind and was hit by fire from the lead fighter. Badly damaged and set on fire, he crashed in a lake. The other three made it to the target and destroyed it along with equipment, intelligence records and the personnel who happened to be inside.

Not shared was what did not happen. Of the bombs released, one scored a direct hit, penetrated and stayed inside the building, but did not explode. The other bombs hit the building, bashed straight through it and out of the other side before exploding on the outside.

The bombs had delay-fuses to allow the Mosquitos to fly clear of the blasts, but the pursuing FW 190s might not have been. The Mosquitos pulled a tight turn at full throttle, and pushed down back into the fjord to escape by the same way they came in with the FW190s in hot pursuit. Having shed their bomb-load and burned off more than two hours fuel on the journey to Oslo, the Mosquitos were light and pulled away from them. Then one of the pursuing FW190s crashed and the pilot was killed. There was speculation that this might have been due to catching the bomb-blasts a few moments earlier. The Luftwaffe fighters, presumably realizing that they would not be able to get close enough to attack, broke off and went home. The remaining three Mosquitos made it back over the North Sea to refuel at RAF Sumburgh at the southernmost tip of the Shetland Islands.

The bomb that refused to explode nevertheless did substantial damage, as did the bombs that exploded on the outside of the building. This caused great excitement amongst the citizens of Oslo. Firstly, there had been a hitting back at the enemy; secondly it was delivered against the HQ of the most hated part of the occupying apparatus, the German secret police or Gestapo; thirdly the aiming had been perfect. Most impressive of all was the apparent ability of the attackers to limit the damage to one individual building in the middle of a row of buildings without hurting big numbers innocent civilians – what today we would call 'collateral damage'. Details of the raid were released to the BBC Home Service which let the nation in on the story in the 6 o'clock News. It was the first public revelation of this pin-point precision capability that was to form an important part of the reputation of Mosquito special operations crews and the aircraft they proudly flew.

A second high profile example happened in January 1943. There were to be two radio broadcasts on Hitler's birthday celebrating ten years of Nazi rule, one to be delivered in the morning by Reichsmarschall Göering. A lone Mosquito bombed the building housing the Berlin broadcasting station knocking it off the air in mid-speech. In the afternoon, the speech to be delivered by Propaganda Minister Joseph Goebbels was similarly disrupted by a second lone Mosquito.

Many raids, where a Mosquito not only had to get itself to the right pin-point but also at exactly the right time were often intelligence-lead. Agents of MI6 and of the Special Operations Executive (SOE) would pass intelligence through their handlers and co-ordinators back to their respective HQs in London, requesting help in an impending crisis. Sometimes it would be about getting someone believed to have been compromised out of 'the field'; maybe a request for technical or operational support; but sometimes it was for something a lot stronger. An example was code-named *Operation Jericho*. At the beginning of 1944, London was informed that a group of over a hundred Frenchmen had been moved to Amiens prison, which had become a kind of concentration camp for French resistance fighters. The British received information from agents that the prisoners were awaiting interrogation and subsequent execution for resistance activities including sabotage and espionage. Furthermore, many of these subversives had been aiding MI19 and other independent networks to arrange the escape, protection and repatriation of Allied aircrew who had been shot down and taken prisoner. For these reasons, any request to help would have received a fair wind from the Royal Air Force. They felt they owed a debt.

Group Captain P.C. Pickard DSO** DFC and a hand-picked navigator Flight Lieutenant J. Broadley DSO, DFC, DFM were chosen to lead a mixed formation for the attack comprising MkIV (Bomber) Mosquitos and Hudson bombers, with top cover escort provided by Hawker Typhoon fighters. Pickard was a star in every sense of the word. His earlier career had tested his nerve flying Westland Lysanders as Commanding Officer of No 161(Special Duties) Sqn, part of the specially formed unit tasked to provide air support to the Special Operations Executive (SOE). The SOE had been formed by Prime Minister Winston Churchill personally. Churchill knew that, like pretty much all wars, the longer it went on the rougher it would become, and this was one of his preparations for when things would get really rough. It was referred to as 'Winston's Department of Ungentlemanly Warfare'. This special operations squadron honed to razor sharpness skills that enabled them to penetrate the airspace of occupied countries, notably France, at dead of night, when the weather permitted and the moon was neither too dark but especially not too light. Being heard is one thing, but if they can see you they can shoot you. Their missions were to insert and extract language-fluent agents who may have been trained in any combination of skills that could include espionage, sabotage, assassination, wireless telegraphy, covert organization, prison escape, cryptography, forgery, breaking and entering, map reading, escape and evasion, resistance to interrogation and disguise. *Operation Jericho* required the local Resistance to get messages into and out of Amiens prison to warn the prisoners what was about to happen so they could be mentally and organizationally prepared to maximize the escape opportunity and to know that once on the outside of the surrounding prison wall the Resistance was waiting to fight pursuing guards and spirit them away to safe houses, hide-outs and escape routes. These were the sort of arcane skills the SOE was trained to furnish.

That should be enough for anybody one might think. But no – Pickard had previously been excused duties to star in the 1940 Crown Film Unit production '*Target For Tonight*' (Coming soon to a cinema near you!), which no doubt had teenage girls, and possibly their mothers, swooning. After a recent conversion to the Mosquito, he had now been selected to lead what would be one of its most intrepid raids. The formation included a photographic Mosquito to capture the whole event. So ironically, Pickard was still to be a film-star, but this time, alas, there would be a much less happy outcome.

The preparations for *Operation Jericho,* took place in conditions of utmost secrecy. The prison was built in the shape of a cross surrounded by wide court-yard with a barracks adjacent or connected to one side to house the guards and other supporting personnel. The entire complex was surrounded by a six metre high wall and a model of the entire complex was unveiled at the briefing for the crews to ponder. One can imagine the crews squatting on their haunches a few feet back from the table on which the model stood trying to create and memorise a mental picture of what they should be seeing in the last few critical moments prior to bomb release as they swept in to attack. The first wave of six Mosquitos was to blow a breach in the outside wall centred on the north-east corner. The second wave was to attack two of the end walls of the prison, firstly to produce shock waves to spring open the prison doors and allow the prisoners to escape; and secondly to destroy the barracks and kill the guards within it to prevent any attempt to pursue, recapture or shoot the escapees. There was a perilously high risk of collision over the target because the two first waves of Mosquitos would be attacking at almost the same time and the attack directions were mutually at right angles. The attack profile called for judgment calls of minutes and seconds, which meant that while the time of the arrival of the whole formation was not critical, the manoeuvre to break the main formation into its three waves, position them, and then co-ordinate the commencement of the bombing runs of each of them, clearly was. The commencement of the second wave was the most critical of all – their timing had to put them over the target as soon as possible after the first wave was clear and their bomb blasts subsided – but absolutely not before! The third wave was the reserve force of Hudson bombers – they were to stand off in case any part of the breeching went awry. In that case presumably Group Captain Pickard would order them in to finish the job in whatever way they could. Finally, a force of resistance fighters was to be assembled covertly to assist the escape and deal with any enemy who had not been killed by the aerial attack. The precise time the attack would begin was to be passed to them at the last moment.

The first attempt was stood down because of atrocious February weather. This was tremendously frustrating as time was clearly of the essence.

On the following day the second attempt met with similar flying conditions, but it was decided that it simply was not possible to put it off any further as the date for the of executions of prisoners could be imminent . Late in the morning on the day of the attack it was still snowing and blowing hard as nineteen Mosquitos carried out a stream take-off from their base in the south of England. During the transit towards the south coast they attempted to form up, which was fiendishly difficult in conditions so bad. The rendezvous with the Typhoons was carried out over the south coast, but only eight out of the slated twelve managed to join up. Then some time after the whole formation had set heading for the target, four of the Mosquitos lost visual contact with the main body and had to return to base.

The projected collision hazard over the target was preceded by two others. The first was during the initial part of the transit, where those four Mosquitos lost visual contact in a close battle formation and had to break away and return to base at RAF Hunsden. The formation that should have comprised thirty-one aircraft was now down to twenty-three.

There was a second – when what was still a very large formation descended to wave-top height to avoid radar detection. In this circumstance the collision risk especially in those conditions was with the sea. The old rule in low-level formation flying 'Never get below your Leader!' could never have been more pertinent.

Approaching the French coast the weather suddenly improved; the relief felt is hard to imagine. The first positioning fix was picked up on coasting in over France. The formation then steered north around Dieppe in a gentle turn towards the east. They were looking forward and trying to pick up a line feature, which they had identified as critical at the planning phase. It was a road lined with tall poplar trees, easy to spot even low and fast. When they saw it and turned right to run along it, the road with its row of poplars would take them directly to Amiens prison, perfectly set up for the first wave to attack the eastern wall.

They identified it and duly turned right dropping down to tree-top height. The formation broke up into the three sections, section two Mosquitos of No 464 Squadron heading for its own attack initial point and the third to carry carry out its planned holding arrangements. Everyone was busy.

After breaking up into the pre-planned waves, command of the first wave of Mosquitos of No 487 Squadron fell to Wing Commander I.S. Smith. Smith now was the lead aircraft of the three Mosquitos he had called into close formation. Their bomb-doors were already open as they visually identified the east wall. All three released their bombs. Smith's bombs scored direct hits on the east wall. Others were close. The north wall was then attacked, one bomb exploding on the prison building. Four Mosquitos of No 464 Sqn attacking in pairs of two attacked the barracks housing the guards. Pickard, flying with '464' orbiting above the mayhem at 500 feet, could see a clear route out of the prison over the rubble and called off further attacks by coded message on voice radio. The reserve Hudson bombers of No 21 Squadron, anxiously waiting, broke off and headed for home. Meanwhile, the Luftwaffe arrived and were engaged in a savage fight by the RAF Typhoons.

The first casualty was a Mosquito of '464', which was hit by flack. The crew, both seriously wounded, managed to crash-land but the navigator did not survive. A Typhoon was shot down near the scene of the raid and another crashed into the Channel on the way home brought down by a snow storm.

Worse news was to come. Pickard's Mosquito was attacked by two Fw 190s, which shot it down killing both crew members. They were first regarded as 'missing', but later the Resistance confirmed their deaths and that their bodies had been taken back to Amiens and buried near to the prison.

Of the 700 prisoners, 258 escaped. It is estimated that 102 were killed during the bombing, some by German machine-gun fire.

It was a raid planned and launched against daunting odds, and a refusal to contemplate doing nothing. London received a message from the Resistance that contained the words, '…we were not able to save all, but thanks to the admirable precision of attack the first bombs blew in nearly all the doors and many prisoners escaped with the help of the civilian population.'

In the intervening years debate about the moral aspects of the raid and about whether it should have been mounted at all have occurred. None of these can detract from the determination and the awesome flying skills shown by the crews on that fatal day. There was worse tragedy to come.

Chapter 14

The Arhus Special Operation

The months after a country is invaded are full of terrifying uncertainty for the conquered people. Where do their loyalties now sit? What will the police, police? What will the government do? What is expected of us? How can we protect our families? How can we stay out of trouble? Who will collaborate? Who will turn spy? How may we minimise the risks of starvation, oppression, suspicion, black market gangsterism, and how do I hide or escape if I am a Jew or a Roma or any other minority out-group? Most people are cowed, not because they are cowards, but because they knew that the rounding up of their loved ones to be put on a guarded train and sent to God knows where had become a distinct possibility. If they did not collaborate.

In filling the vacuum of the governance of Denmark, Aarhus University was selected by the Gestapo as its head quarters. The secret state police operated with no mercy.

There were a few though. Maybe they and their families had already been rent asunder. Maybe they had already been separated and left behind filled with tears. And some, outraged by the violation, just wanted to hit back to destruction. Wracked with hatred and determination, they went underground filled with a terrible resolve.

The Gestapo selected Aarhus University to site its new headquarters for control of Denmark. British espionage within

Denmark did not get off to a good start. Initial infiltrations in1943 failed and resulted in arrest and torture by the Gestapo, which broke agents revealing information about contacts and networks. They were duly shot or rounded up and imprisoned. There were many betrayals under extreme pressure.

The British war authorities, at a loss about Denmark, and as a matter of extreme need, sought the assistance of the clandestine Special Operations Executive (SOE). This was a matter of some delicacy as the British Army rightly suspected they might not be able to control them. However, they had certain attractive advantages.

Firstly, they were the most fearsomely trained covert units available. Neither the Germans nor the Americans could field anything directly comparable with them in the early part of the war.

Secondly, their way of operating was deniable – anything they did would never come back to stick to you. The SOE knew how not to exist.

Thirdly, they cost little money – they were trained in misappropriation. At an Annual Dinner of the British Special Forces Club held some years ago at the Imperial War Museum in London they were acknowledged by their Patron, Anne, The Princess Royal, as the only unit in the British armed forces that ended the war in profit.

Fourthly – in the tradition of T.E.Lawrence, they believed in, and were trained for doing what the enemy believed was not possible.

Fifthly – Churchill would approve it – they were his baby. 'Winston's Department of Ungentlemanly Warfare' generally got what they asked for.

The SOE liked to get up close and personal. They sent messages to high ranking officers of the *Wehrmacht*, and particularly the high-ups in the *Schutzstaffel* (SS) and the *Geheime Staatspolizei* (Gestapo). The messages were not threatening messages delivered by wireless or spies. They were delivered by killing senior officers who thought they were immune. *Operation Anthropoid* was an example. Four Czech members of the SOE, trained in Britain, were parachuted into Czechoslovakia in an RAF aircraft. The drop was not very accurate. After time in hiding and travelling at night, they managed to infiltrate the city, and eventually ambushed Reinhard Heydrich, the SS commandant of Czechoslovakia responsible for the rounding up and murder of Jews and others regarded as *untersmench (*sub-human*)*. Heydrich made the basic mistake of not varying the time and route his driver took carrying him between his quarters and his HQ. He had also made himself vulnerable by travelling in an open-topped vehicle. As the car rounded a corner the lead SOE agent stepped into the open with a Sten submachine gun levelled and pulled the trigger. The gun jammed before the job was done, not unusual for the Sten, but with great speed and presence of mind, he tossed aside the gun, pulled a grenade out of his pocket and lobbed it into the car. They explosion did not immediately kill Heydrich; he was seriously injured and died some days later. The grenade had blasted the horse hair that filled the seats of his car into his body, and he died of blood poisoning. A massive manhunt was mounted to corner and arrest the attackers. They were eventually hunted down in Prague, were cornered in a church and after holding out for as long as they could, they swallowed cyanide pills and shot themselves in the mouth to avoid capture. But the message was clear. The SOE will find you and will kill you.

Later in the war the SOE presented a plan to assassinate Adolf Hitler (*Operation Foxley*). By that time in the war their field-craft was so skilled and practised that the plan was feasible. It was cancelled by Winston Churchill on the grounds that Hitler was making so many strategic mistakes that the Allies needed him to remain alive.

It is well known that there was an important connection between the SOE and the RAF through the career of an aircraft called the Westland Lysander. The Lysander was designed as an Army Cooperation aircraft; high winged giving maximum visibility of the ground from the cockpit; very low landing speed; rugged undercarriage able to cope with landing on unprepared airstrips indicated by masked torchlight. It did sterling work carrying secret agents in and out of occupied Europe. A powerful and detailed account of these operations is contained in *We Landed by Moonlight* first published in 1978 by Group Captain Hugh Verity DSO*, DFC (RAF Rtd.) He himself flew many of the missions described so graphically in his book, and one of his fellow officers on the newly formed No 161(Special Duties) Sqn was none other than Wing Commander 'Pick' Pickard, star of the film *Target for Tonight*, and whom, it will be remembered, was killed leading the raid on Amiens prison. Prior to this, 'Pick' had been a Special Duties (SD) pilot similarly involved in the establishment and development of procedure for secret Lysander flights delivering and extracting agents and equipment in occupied Europe, before transferring to the Mosquito. Hugh Verity also became a member of the Special Forces Club based in London, formed by members of the SOE immediately after the war. Both he and 'Pick' Pickard are still remembered and honoured by the Special Forces Club.

The Mosquito in its various guises became the aircraft of choice when something more destructive had to be done, but which required as close to pinpoint accuracy as possible to limit collateral damage. This might require only one aircraft, or four as in the Oslo raid, or bigger formations such as Amiens prison. To answer this request needed a big formation. To support the Nazi occupation of Denmark, the Gestapo based themselves in the buildings of the University of Aarhus. They deployed their staff into two dormitories. The resistance movement in Denmark was desperate. A message was sent to British Intelligence.

'Resistance in Jutland about to be torn apart by Gestapo. Insist attacks on Aarhus University dormitories Four and Five'.

The Gestapo was the target.

The RAF having carried out these sort of requests before, did have experience on which to draw. The Oslo Raid had also been an attack on a Gestapo HQ by four Mosquitos, and at the beginning of 1944 *Operation Jericho,* the bigger Mosquito attack to release Resistance prisoners from Amiens prison. No 544 Sqn was tasked for a photo-reconnasisance mission using PR Mosquitos. They did a good job.

Members of The Special Operations Executive were invited to the briefing of the Mosquito crews. By luck or by evil design, the Gestapo had sited themselves near hospitals. There would be a need for pinpoint accuracy to avoid killing civilians.The RAF Mosquito crews waited for their instructions, hoping they were going to be do-able. A full-sized training mock up was constructed in the open and the crews did practice runs on it, still in the dark as to any details.

When they reported to Ops for briefing, the briefing concerned itself as always with the weather over Europe and the crews were told that this was to be a large attack by a force of twenty-four Mk VI Mosquito fighter-bombers. They were shown the photographs of the targets. They were told the direction that they should lead the attack flights in and why. The attack would be delivered in four waves in quick succession, the first dropping high explosive weapons with delay time fuses to smash up the building and cause maximum destruction without damaging the delivering aircraft. Immediately afterwards subsequent waves would set the buildings and contents on fire with incendiary bombs to destroy as many files and records as possible.

They would have a squadron of P51 Mustang fighters above them in case the formation was detected and Luftwaffe fighters were scrambled to intercept. Whatever was happening at roof-top level it was anticipated there would be a dogfight a few thousand feet above it.

The date was 31 October 1944. Transit from the base in Norfolk to coasting in over the west coat of Denmark took about two hours almost all of it over the North Sea. The wait, faced with such a challenge in front of them, was tense. There was no conversation between the planes until they broke out of formation into four smaller groups shortly after they had coasted in over Jutland. All four waves made for the initial point (IP), a conspicuous lake, from which they would set heading when their turn came to run in to the target. The attacking run would take about three minutes. By this time there would be no doubt to subsequent waves where the target was because the first wave would have dropped high explosive weapons and it should be marked by a tower of smoke and debris. The following waves were armed with incendiary bombs to set the wrecked buildings ablaze.

The Germans were alert and got wind of the attack within 15 minutes of it approaching the coast. The P51 Mustangs broke away and flew straight towards the nearby Luftwaffe base to engage their fighters if, and as, they took off. As each wave of Mosquitos went in, the risk of flak damage increased as surprise dwindled. A cruiser in Aarhus harbour commenced firing at the last two waves of incendiary carriers. The third wave got through unscathed but a Mosquito in the last wave took considerable damage. The Mosquito force, now weighing much less than when they left their base, tore off towards the west and home in good order. There was no interference from the Luftwaffe, so the Mustang pilots had done their job.

The damaged Mosquito was still flyable (just) and, in accordance with the contingency plan, turned north over the Kattegat. One of his squadron broke off to escort until it was clear of the Kattegat and fighter interference.

The crew pulled off a landing in neutral Sweden, and destroyed their aircraft.

They were arrested by the Swedish police and imprisoned. All the other aircraft had a rather more relaxed return than outbound and landed back in Britain just over two hours after the raid.

The estimates of those killed varied. The Germans admitted forty killed of which well over half were SS officers seconded to the Gestapo. The rest were administrative staff. In addition, eighteen German soldiers perished. Important leaders escaped to neutral Sweden. A number of Danes were killed by a bomb that that had gone astray. The important objective of the destruction of the Gestapo files and records that were the administrative basis of persecution in Denmark was achieved. To recover from this serious setback to their operations the Gestapo not only replaced but doubled the number of their staff and operatives. They had to start their malign efforts again from scratch and time for them was running out.

Chapter 15

The Hague and Bonneuil Special Operations and Operation Carthage

Special operations raiding, directed at perpetrators of Nazi oppression within occupied Europe continued. In April 1944 a request was received from the Dutch underground to attack a former art gallery in The Hague, now used as a Gestapo Headquarters (HQ). As was usual the building was now full of documents and identification details of every citizen in Holland. The underground wanted this building razed to the ground, but as usual it was part of a larger building complex. Pin-point accuracy was again necessary to avoid hitting any of the surrounding buildings because that would result in a heavy toll of civilian casualties. The building was defended by anti-aircraft guns. RAF planners got busy with designing the attack profile, a scale model was built, and a practice flying program produced. Training began.

On the selected day, six FBVI fighter-bomber Mosquitos took off from Hampshire escorted by Spitfires. The formation crossed the North Sea at wave-top height and on coasting in re-formed into three pairs. There were High Explosive (HE) bombs in the bomb bays to blast open the buildings, plus incendiaries to burn as many of the records within them as possible. The two following pairs did a circuit around Lake Gouda waiting for the delay fuses in the first wave bombs to explode, then turned in to attack too. The Spitfires climbed and held off watching to see if there would be any attempt to oppose them from the air and if so to engage. The raid went to plan and the six Fighter-bombers broke away low level again for the home run escorted by the Spitfires. The Gestapo HQ had been completely destroyed and its contents of paper records with it. There was little damage to surrounding property, but there had been 130 civilian casualties, sixty-one of them killed. The daylight pin-point attacks against specific buildings were not commonplace, but they continued to be requested by resistance leaders through the SOE despite the risks, and continued to be flown.

Sometimes it got very personal. Now that D-Day had come and gone, Allied regular troops were engaging with German troops on a scale that had not happened since Dunkirk, and that inevitably brought them into contact with the SS and others prepared to engage in viciousness. On one such occasion an order went out to assemble a Mosquito force for a special mission against a Gestapo barracks at Bonneuil that housed a unit which had murdered some British paratroopers in a town square. These were members of the newly formed Special Air Service (SAS) who had been attacked by a German unit; while some escaped, others were captured and murdered in a particularly brutal fashion, clubbed to death in front of villagers forced at gun point to witness the atrocity. The escapees radioed the incident back to Britain having identified the unit and the barracks in which they were billeted. The barracks were a collection of buildings. The Mosquitos were to destroy the buildings at evening meal time while this unit was at supper. Once again, the target buildings were close to a village that was not to be struck so the highest possible accuracy was required. This raid took place and approximately 150 soldiers of the offending unit were killed. However, some survived, but not for long.

The SAS was quick to learn the potency of being able to whistle up a squadron of 'Mossie' fighter-bombers and direct them on to targets; so when, from their hides and secret patrols, they detected a force of 2-3000 troops assembling at a barracks near Poitiers, they asked for help. The RAF duly obliged by sending over twenty-four Mk VI fighter-bombers with an escort of Mustangs providing top cover to neutralise the threat. A little while later the SAS were tipped off that the survivors of the unit attacked at Bonneuil, which had murdered their prisoners, was regrouping in another SS police barracks. They reported its position to London, and back came the RAF this time with 23 FB VI fighter-bombers, protected overhead by Mustang fighters, to finish the job.

The litany of special operations goes on, from attacks directly upon the enemy's unspeakable state persecution activities, to clandestine flights to and from neutral countries, to the development of revolutionary new weapons and more. The ability to attack a 'pin-point' target is obviously an exaggeration but there is no doubt that these operations were conducted with great skill and bravery. There is no point trying to ignore, however, that they were very dangerous to anyone involved in flying them, or who might have been unfortunate enough to have been under them when they were delivered. The question of 'collateral' damage as we would now call it must have been ever present in the minds of the planners and the crews. It is a safe bet that if one keeps on taking big risks that one day it will go wrong, and that is what happened most grievously in *Operation Carthage*, the raid on the Copenhagen Head office of the Shell Company, which became known as the Shellhus Raid.

It was the same set up – the local resistance needed to strike back hard at the ability of the Gestapo to break their will to fight by using torture, executions, and threatening the lives of friends and family. They needed the local HQ bombed out. The desperate requests for a pin-point attack started to come in. The Royal Air Force tactical planners looked at it and declined on the grounds that it would be so difficult that the risk was unacceptable. The building, called the Shellhus because it was built for the Danish head offices of the international oil company, was now occupied by the Gestapo. It was located down town surrounded by other buildings. It was defended by anti-aircraft guns. The calculation was that possibly as many as 300 civilians would be killed and losses to the attacking squadrons would also be high. The British Air Board, who had to agree to any attacks on such sensitive targets within Allied albeit enemy occupied countries, ducked the decision and handed it back to the RAF. Air Vice Marshall (AVM) Embry, commanded 140 Group known as 'The Gestapo Hunters'; they were the obvious choice for this raid. Embry was unhappy with being put on the spot and demanded that the final order would only be given on the best of intelligence assessment, and this would have to be carried out and relayed back by an operative on the ground in Copenhagen. It was already known that the top floor of the building had been converted to prison cells and had been filled with resistance captives, an obvious human shield to deter against air attack.

However, the Special Operations Executive (SOE) had been working for some time with the Danish resistance, former Danish intelligence operatives, and British Intelligence. Coordination between these organizations had produced a lot of information. This allowed a scale model of the target area to be constructed. The

British picked Ole Lippmann, a senior Danish intelligence officer to make the assessment. His assessment included communication interceptions that indicated that the enemy was now making a maximum effort to destroy the Danish resistance. This enforced pacification would allow the release of 100,000 of their troops to be thrown in to fight the Allied advance through Europe towards the Battle of the Rhine. The proposal therefore assumed enormous strategic importance. Embry, who was in the habit of flying on these missions himself under a pseudonym to try to conceal his identity in the event of capture, with considerable worry, relented. It was the best assessment available and, whatever the risks, the strategic requirement to prevent the re-deployment of those troops must have played a critical part in the decision.

In March 1945 eighteen of the redoubtable Mk VI Mosquito fighter-bombers were assembled in Norfolk. This made it not only a special operation but also an extreme range operation with all the extra risk that this entailed. The whole operation would take a round trip of five hours or more depending on the weather conditions and particularly the wind direction and strength. The briefing commenced during the following day in conditions of strict secrecy and crews were confined overnight, under armed guard, to prevent them mixing with other personnel. Briefing details were finalised on the morning of the following day so that crews could update on the latest weather, wind vectors and the latest intelligence. Accurate wind vectors were critical to making the round trip without running out of fuel. Air Vice Marshal Embry turned up to fly leading the first wave. The formation contained two Mk IV bomber variant Mosquitos that had been modified to take filming cameras to record the raid.

The MK IV bombers were flown by crews and cameramen of the Film Production Unit (FPU). They retained their offensive capability however and carried 500lb bombs. The first was to fly in the first wave both bombing and filming at the same time; the second with the third and final wave, also bombing as well as filming the results of the first two waves. Special ops were in the habit of including such filming crews and their job was even more dangerous. Once a fighter-bomber or a bomber had released their ordnance (or 'stores' as they were then called) the crew was free to make their escape from the scene. This could be via a re-formation into groups for mutual cover in case of trouble on the way home; or the briefing might be to make their own way home as singletons, especially if it was going to get dark during the transit. Either way, once the bombs had gone you didn't hang about, but removed yourself from the battle area to reduce the risk to your aircraft and crew and to avoid getting in the way of other crews who still had business being there.

Not so the film crews. Like the formation leader, they had to loiter as long as necessary at the scene of the attack, in continuing danger, to film what had happened for analysis when they recovered to base. Keeping the cameras rolling while on a bombing run yourself, ignoring the ground fire bursting around you, and then having the possibility of more runs over the target in the aftermath must have taken a lot of self-control. The films of these bombing attacks are riveting.

During the transit the Mosquitos would be formed into three waves. Mustang fighters would weave around and above each wave. The Mustang was the only single engine fighter on the Allied side with the range to escort them out, engage the Luftwaffe if it turned up, then fly all the way back to England. As well as the close-in escorts, a flight of Mustangs was tasked to sweep the airspace north of the formation's track to effect early warning and disrupt any attempt to intercept by the Luftwaffe.

The morning takeoff and forming up went without incident but the weather was stormy and unhelpful. The wind was around 50 knots (58mph), gusting and turbulent. Height was reduced to as close to 50 feet as possible to avoid radar detection, but the buffeting made controlling their aircraft hard work for the pilots even in loose formation. The lead navigators were starved of fixing information once they had coasted out from Norfolk with nearly 400 miles of a long featureless North Sea in front of them. Any help they sought from radio navigation aids would have been diminished by their ultra-low height. The other navigators can be imagined watching the fuel flows like a hawk and with any significant shift in wind vector would immediately re-do the round-trip fuel recalculations to confirm that they still had enough. Doing such a recalculation and coming out with a no-go result is not a good feeling when one is a long way out over the sea. One blessing was that such low-level proximity of the aircraft to the sea surface would have obviated any icing problems; but there is no such thing as a free lunch and it came at the cost of sea salt whipped up by the storm that began to coat the windscreens and impede forward visibility. Another consequence was that three Mustangs suffered bird-strike damage and had to abort. Notwithstanding all this, the formation made a landfall at the planned coasting-in point.

The three waves now had to be separated so they could be co-ordinated and didn't get in one another's way. The selected Initial Point (IP) was a conspicuous lake some miles away from which a planned heading could be steered to take them towards the target area in Copenhagen until it was visually identified. The second wave was to make one gentle turn around the lake for time separation before heading off towards the target itself. The third wave made two circuits of the lake for the same reason and then also set heading. The time separation between each wave was approximately two minutes. The critical calculation in flying these circuits was that the time fuses in the bombs would detonate with sufficient delay to allow the delivering aircraft to get clear of the blast, but not with sufficient to cause a threat to the next incoming wave. That aspect of it worked. Some of the escorting Mustangs dived to 2,000 feet above the attackers to spot any flak and destroy the offending batteries. The rest stayed thousands of feet above with their eyes peeled for the Luftwaffe.

The first wave pilots adjusted spacing towards close formation, opened the bomb doors, throttles increasing power to increase speed and keep station. With about half a mile to run, inexplicably, an aircraft in the second rank of the wave started to lose height. The formation was at just above roof-top level so there was absolutely no room for that kind deviation. The aircraft struck a pylon smashing part of a wing causing irrecoverable damage. Out of control, the aircraft reared up to the vertical and rolled, nearly causing a mid-air collision. The attack however, was driven through with the bombs of the remaining aircraft in the first wave aiming at the Shellhus lower floors in an attempt to avoid killing the imprisoned resistance fighters on the top floor. The building was struck at street level. The bombs of the stricken Mosquito came off, one striking a building and killing eight people. The delayed action bombs that hit the Shellhaus exploded. It was what happened to that doomed aircraft that caused the main tragedy. It crashed in close proximity to a school full of children and teaching staff. The fuel tanks exploded on impact causing a huge pall of smoke to rush upwards into the air. The first wave dispersed and set heading for home.

The second wave was now close. The operation of this form of low-level attack assumed that the first wave would hit the target which would start the destruction. The column of smoke, fire and debris would act as a target marker for the following waves. The problem was that, due to the crashed Mosquito, there were now two columns of dust and smoke. In the few seconds that followed, the crews in the following wave had to make a decision – which was the right column of smoke to bomb? One aircraft made the wrong decision and bombed the school with great loss of life.

Late in the attack, the flak started. The third wave of attack was now unnecessary. Some aircraft circled trying to work out what was what. In came the third wave and the confusion increased. Four of its Mosquitos re-bombed the school. One third wave pilot reported that after much confusion and realigning he positively identified the Shellhaus, which was shattered and on fire, concluded that there was nothing more to be gained, and broke off for home.

The journey home was not pleasant. The crews knew this had not gone well, but not all knew just how badly. Some were hit by ground fire and all of them had to spend around two hours not knowing if they were going to run out of fuel and be forced to ditch in a stormy North Sea where the chances of rescue would be poor. One crew, fuel down to last knockings, coasted in just in time and landed at an aerodrome but they didn't have a clue which. They were approached by armed Military Police and invited to accompany them to explain themselves. One Mosquito that had been hit by fire from a German cruiser moored in Copenhagen harbour, tried to get to Sweden but didn't make it. It is thought they ditched in the sea short of the coast, but they were never found. One Mosquito had an engine damaged and made it back to England on one engine. One Mustang was hit over the target and ditched in a fjord in Jutland. The pilot was captured and imprisoned.

In all, four Mosquitos, two Mustangs and nine aircrew were lost in the raid.

Of prisoners in the Shellhaus eighteen escaped and eight were killed.

At the Jeanne D'Arc school in Copenhagen, eighty-six children and sixteen adults were killed; sixty-seven children and thirty-five adults were wounded.

The single bomb that came from the first stricken Mosquito killed eight adults.

The Germans lost 102 Gestapo and Danish collaborators killed; and their records and files were destroyed.

The Danish underground thanked the RAF and regarded the operation as a success albeit at a very high price.

Only one more such operation was flown. In April, intelligence reports came in informing that a building on the edge of Odense in southern Denmark had been appropriated as a Gestapo HQ. Six Mosquitos went out on the operation and destroyed the building.

The Allies continued to advance on all fronts and, shortly after this raid, Denmark was liberated. For the brave, long-suffering Danish people, the agony was over.

Chapter 16

Tactical or Strategic?

In the aftermath of the Battle of Britain, Winston Churchill uttered his most famous words.

'Never, in the field of human conflict, has so much been owed by so many to so few!'

But true as these words were, they were not as significant as the words he uttered a short time later.

'The fighters are our salvation; but only the bombers can win us the war!'

What does 'strategic' mean? It is a much misused word. It is used as a synonym for a plan. It is a plan, but it is a plan of a very specific kind. It is a plan that embraces the likelihood that its implementation will be opposed by a rival, a competitor, an enemy. It has to 'second guess' what a rival will do immediately after your first move, or even in a pre-emptive move, the details of how that might weaken your own position, how you might minimise this, and how you might tempt your rival into making a wrong move. The science of strategy is 'Game Theory'. Developed in the first half of the last century, one of its first military applications was by the Royal Navy as a way of predicting the positioning of U-boat packs in the North Atlantic.

Strategic decisions, by definition, can have enormous consequences. In the First World War the British Admiralty subscribed to a fanciful doctrine which declared that battleships could be made faster and more manoeuvrable by shedding armour plating. This could be done safely because speed and manoeuvrability would make them very difficult to target by an enemy. The battlecruiser was born. However, it was a mistaken doctrine and, as the Battle of Jutland demonstrated, they were rendered highly vulnerable. The German High Seas Fleet lost over 2,500 men, but the British losses were over 6,000 sailors. The majority died in the three vulnerable RN battlecruisers that were sunk causing Admiral Beatty to complain that there seemed to be 'something wrong with our bloody ships today'.

He was not wrong – the weakness had been mistakenly but deliberately designed in.

The Battle of Jutland emphasised the distinction in another way – it was a clear tactical victory for the German navy, but it was a strategic win for the Royal Navy. The German High Seas Fleet retired back to its bases and did not put to sea again in any numbers until the end of the war. At cessation of hostilities, it was ordered to set sail for Scapa Flow in the Orkney Isles, where it would surrender to the British authorities. The fleet was eventually scuppered at anchor in a last act of defiance. The ships settled on the bottom and are still visible from an aircraft to this day. The Royal Navy had preserved its dominance of the shipping lanes and blockaded the Skaggerak in an attempt to starve Germany into surrender. An American newspaper carried the comment that the German navy had assaulted its gaoler but was still in gaol.

When, after all the training with bouncing bombs, No 168 Sqn RAF was denied permission to seek to attack the *Tirpitz* in a Norwegian fjord, this was a strategic decision. The *Kriegsmarine* was employing a strategic policy known as 'fleet in being', which meant that if you had powerful top-class vessels in your navy you didn't actually have to put to sea. You just had to have the ability to do so to tie up an enemy's navy, itself at sea, to protect your convoys in the event of an aggressive break out. However valuable the tactical gain in sinking such a formidable battleship, there were other considerations that weighed heavily in the balance.

No 617 Squadron had carried out its 'dam-busting' raids in the black of night. Their targets, the dams providing hydro-electric power and water to the industries involved in German war production, were big. The aerial mines that the Lancaster bombers carried were critical in that they had to explode against the dam wall about thirty feet below the surface of the reservoir above the dam to maximize disruptive effect. That was achieved by dropping them from a precise height at a precise speed. The height was achieved by converging two beams from searchlights under the fuselage of the aircraft, which had the huge disadvantage that the anti-aircraft batteries defending the dams could see them coming. However, they could not observe the precise tactics and the delivery mechanisms, nor, it was hoped, the fact they had been attacked by a revolutionary new weapon – a mine that could bounce off the surface of water, hit the face of a structure, roll itself down the face of the dam to an optimum depth and then be exploded by an pressure pistol.

No 618 Squadron however had a greater problem – daylight. The *Tirpitz*, vast as it was as a vessel, was tiny compared to a reservoir dam. It was hidden by cloud or by smoke generators and inside a kink in a Norwegian fjord. There were two schools of thought doing the rounds. One was a medium-level attack by Lancasters. The other was a low-level strike by a flight of specially modified Mosquitos carrying two, much smaller, bounding bombs each. The Mosquitos of No 618 Squadron were modified to take these weapons and the crews trained to use them, all in complete secrecy. However, while the *Tirpitz* was located by a Mk 1 photo-reconnaissance Mosquito, it was eventually sunk by Lancasters of No 617 Squadron on 13 November 1944 after massive air and sea operations proved unfruitful. Which begs the obvious question of why the Mosquitos and crews of No 618 Squadron, trained and equipped at great expense, were never deployed. As tactically essential as sinking the *Tirpitz* was, it was overridden by a strategic need to keep the development of the bouncing bomb secret. It was only taken after all the effort and money had been expended. Could it be that this issue had not been thought through? Or was it an attempt to keep options open? The attack could not have taken place under the cover or darkness. In broad daylight there would have been hundreds of eye witnesses and hundreds of feet of cine film that could be inspected to reveal exactly how the attack had been achieved. Once the cat was out of the bag it could be copied and this could now propose a real threat to the Royal Navy. The vessels of the RN were not a 'fleet in being' – it was a fleet at sea and a fleet in harbour for maintenance and thereby potentially vulnerable to a weapon as effective as Barnes Wallace's had proved to be. It was suppressed on these grounds.

Tactics are grounded in battle area fighting. Strategy is grounded in a much wider environment. Strategy is fought by factory outputs, and the combatants' ability to build it faster than the enemy and destroy the enemy's capacity while defending its own. Fighter aircraft of those days were essentially tactical and defensive. Bombers, however, can be used strategically in a way that would have been unusual with fighters. The targets can be industrial rather than merely military, railway trains rather than tanks, bridges rather than block-houses, oil tankers rather than cruisers, factories rather than naval dockyards, dams rather than airfields, populations rather than armies. The aim tends not to be fighting the armed forces of the enemy, but preventing and minimising the ability of the enemy to field armed forces in the first place.

As a photo-reconnaissance aircraft the Mosquito's role was strategic – its business was information and intelligence gathering – not dropping ordinance. As a bomber it could be used both strategically and tactically as epitomized by its employment as a Path-Finder and a Target marker for the main bomber force. As a day and night fighter its use was tactical and defensive. As an Intruder and a Ranger it was tactical and offensive. When it directly targeted personnel such as the GESTAPO it was being used for special operations. In its deployment to destroy merchant shipping in the Bay of Biscay, the North Atlantic and the North Sea it was being used strategically. Deployed by RAF Coastal Command to destroy *Kriegsmarine* ships and especially U-boats one might argue it either way. It appeared that the Mosquito could achieve every role that was asked of it. Few, if any, of the warplanes of the Second World War could make such a claim.

Acknowledgments

I am grateful to my sister, Kathryn Elizabeth Pearson for her eagle-eyed critiques.

I am also indebted to the Special Forces Club for their kind permission to include a copy of the painting "The Gestapo Hunters" by Michael Steele Morgan, which hangs in their premises in London.